For Ardis,
companion in the wilderness,
and David,
our link with the future.

*Philip Hyde*

For the spirit
of the rocks.

*Edward Abbey*

# Slickrock

Endangered Canyons of the Southwest

Words by Edward Abbey
Photographs and commentary by Philip Hyde

Sierra Club/Charles Scribner's Sons • New York

All photographs Copyright ©1971 by Philip
Hyde. Used by permission.

Designed and produced in New York by
Charles Curtis, Inc.

Filmset in 14 point Trump by
TypoGraphic Innovations, Inc., New York.

Printed and bound in Italy by Mondadori
Editore, Verona. Library of Congress Catalogue
Card No. 73-163897. International Standard
Book No. ISBN 87156-051-8.

# Contents

# Introduction

For my part, this book began on a windswept airstrip near Moab, Utah. Phil Hyde was the first to greet us, and that was appropriate because it was Hyde who really started it all. Hyde and his camera. Hyde and the maps on the walls of his studio in Taylorsville, California. Hyde and the red lines, blue lines, yellow lines, green lines traced across those seven years and all the brown contour lines of the up-and-down country, each felt-tip color designating yet another trek with tripod across the most photogenic landscape east or west of Eden. Hyde and his deep loathing for the slobs who would ruin it all. Hyde. With a sandstone beard and a red bandana. So it began.

An ersatz Hobokenite was there, too: Ed Abbey, home again in his solitary desert, hatless in the hot May sun. No stranger to the canyons, either—though until three weeks earlier, a total stranger to Phil Hyde. Abbey and a friend had been exploring a small canyon near the Doll's House. Friend, scrambling solo, encounters man with tripod on the rimrock. Cameraman explains he is doing a book. Funny, Friend says, so is my buddy. Cameraman asks identity of buddy. Ed Abbey, says Friend. Funny, says Cameraman, *same book*. Friend hollers down canyon: Hey, Ed. Guy up here says you're collaborators. Abbey scrambles up. Ed Abbey, says Friend, meet Doctor Hyde.

So perhaps the wilderness is not so large a world after all. Sad, isn't it?

From the Moab airport we drove south toward the canyons in a Sunkist-colored four-wheel-drive wagon. We were six: Kent Frost of Monticello, driver and guide, river-runner, cowboy, prospector, campfire raconteur; beside him, Hyde and August Frugé of Berkeley, chairman of the Sierra Club's publications committee; in back, our art director from New York, Charles Curtis, and six-foot-three Abbey, fish-hooked between his new straw sombrero (ninety-eight cents at the Moab drugstore) and Frost's aisle seat (an overturned washbucket). Crowded? Who cared? We were heading for canyon country. To get *uncrowded*.

The idea of the expedition was to bring together—on location, as it were—the writer, photographer, designer and editor, so that collectively we might begin to share in conceiving the book even before the first photograph was selected or the first word written. We sought, in particular, a symbiosis of word and picture. Each day in the Sunkist wagon, or tramping cross country between jeep trails (where the sources of inspiration are more accessible), we discussed a dozen different ways one might distill that incredible country into 144 pages. But in the late afternoons, after making camp, everyone went his own way. Frugé stalked the wild spring desert flowers then in bloom. Frost scouted the canyon ledges for Indian ruins. The two New Yorkers debated the possible proximity of scorpions and rattlesnakes. Hyde slipped away with his camera. And Abbey scrambled to the apex of some nearby arch, there to meditate, his straw sombrero pulled low against the setting sun. In the evenings, around the campfire, we sought consensus on a title for the book.

Yet how does one begin to describe such country, much less crystallize its essence in twenty-five words or less? Canyons, mesas, buttes, desert and glen, yucca and cottonwood, deep skies and twisting washes—and of each a hundred, a thousand variations. Can you call it wilderness, someone wondered by the firelight, if the wildest parts of it are unprotected and already threatened by the fallout from power plants and by the dotted lines (oh, so much straighter than Phil Hyde's) on the one-dimensional maps of the highway engineers?

From the smoky side of the campfire came a suggestion. "For the title," said Abbey, "how about *Help Stamp Out Rape?*"

Tabled. But at least it was a beginning.

There was no consensus on a title, either at Beef Basin, or Chesler Park or at the final camp in lovely Lavender Canyon. Still, one word (or rather two of them put together) kept recurring. *Slickrock.* Everyone tried adding other words before and after it. Nothing worked. We retired to our sleeping bags.

In the middle of the night I awoke. The moon was floating overhead on a river of sky between the high canyon walls. Light sparkled

Pot Canyon off the Escalante (before flooding)

Horse Canyon in The Maze

along the vertical edges of the Cedar Mesa sandstone. Sandstone polished by time and primordial forces. Smooth. Slick...

Breaking camp the next morning, I asked Kent Frost about the origin of the term slickrock. "It isn't necessarily limited to any particular geological formation?" I asked.

"No, it isn't," Kent said. "It's more like the whole country is slickrock, even where it's crumbling away." So it became *Slickrock*.

As the book moved into the final stages of production, I realized that the symbiosis of words and pictures we had sought was, after all, even more elusive than the title had been. Edward Abbey is a writer, not a caption-manufacturer; and Philip Hyde is a photographer, not an illustrator-of-text. Each sees his world differently, and uses different tools to express his own experience of that world. And this is as it should be. Thus, the book comes in two sections: *Abbey's part* and *Hyde's commentary*, two distinct views of the slickrock country. Complementary, yes; symbiotic, no.

My memories of the photographer and the writer on that trip are likewise distinct. I remember Hyde and his tripod atop a sandstone needle in Chesler Park, silhouetted at sunset against the tangerine sky. I remember Hyde's head disappearing under the black hood of his view camera, then the long immobility as he waited for the right light, the exact moment. And when it came, from under the hood Hyde brought forth a piercing yodel. A direct response to a moment in time. A strong, colorful statement. Just *like* Hyde.

I remember Abbey in a sandstone window overlooking a maze of canyons that wind off toward the deep gorge of the Colorado River. He was chewing on a blade of grass and the sombrero was low again in observance of sundown. *Just like Abbey.* Darkness was coming on fast. Time to return to camp. Abbey removed the hat and, holding it level, slowly extended his arm toward the big river. Though it struck me as an unusual gesture, it was at once natural and moving. Abbey, saluting the slickrock with that silly sombrero, reaching out to bless the stark chiseled bounties of that wild beyond.—*John G. Mitchell/editor, Sierra Club Books*

# Preface

On Earth Day, 1970, the day some people believe was a turning point in the environmental crisis in America, my family and I walked into the Escalante canyons. It seemed a good place to be on Earth Day—close to the Earth, yet on a mission that could possibly contribute something to help in the crisis. We need wild country where the natural systems that support life may be drawn upon, not only for inspiration but to teach us what we need to know when our cultivated systems go awry.

It was the imitation of natural life-support systems that made possible manned space flights to the moon. The imitation, though only a pale, cramped shadow of Spaceship Earth, nevertheless put men into space far enough to give us all a vitally important view of Earth—a view of Earth as a still beautiful brown, white, and blue ball, tiny in the immensity of space, the only known oasis in a dark, cold, possibly lifeless desert of endless space.

Perhaps that view can help even the most insensitive of us to understand finally that Earth is not a platform on which to build the ever-multiplying human ant hill, or a raw material to manufacture into something else that seems momentarily more useful.

With this view of Earth as our home, as the very womb that nurtured mankind from his earliest beginnings, perhaps we can begin, finally, with love and care, to protect its life-supporting systems—and so make it possible for mankind to go on being its tenants—not just for this generation, but "for all the days," as the five-year-old who lives with us is fond of saying.

The focus of this book is on a part of Earth that is still almost as it was before man began to tinker with the land—a piece of wild Utah that ought to remain that way. Telling thousands about it—to get their help in what must be a prolonged struggle to keep it wild—is a calculated risk. Canyon country is fragile, vulnerable to human erosion. I have some hesitation in showing more people its delightful beauty—hesitation born of the fear that this place, like so many others of great beauty in our country, might be loved to death, even before being developed to death.

So, if our book moves you to visit the place yourself sometime, first make sure you add your voice to those seeking its protection. And if you do go there, tread lightly, pack out your garbage, leave so little trace of your passing through that the next one in may fancy himself almost the first to come there. And above all, don't ask that your getting there be eased. It is easy enough now if you can walk, as our five-year-old has proved. If you could get there any other way, you would not find its essence and wildness.

This piece of country has three related parts. First, in order of wildness is the Escalante basin, a remnant of the once-vast wilderness of Glen Canyon, much of which has been reduced by the reservoir behind Glen Canyon Dam to a water playground of the motor culture. Adjoining the Escalante on the northeast is Waterpocket Fold country. The third part, loosely referred to as Canyonlands, is separated from the first two by the last discovered mountain range in the contiguous United States, the Henry Mountains. Alas, the Henrys were wild in the recent memory of many who know that country, but they are now laced with roads whose engineering standards are almost yearly elevated.

Canyonlands includes what is now Canyonlands National Park, as well as the wild, seldom visited area on the west side of the rivers that should be a wilderness part of the park, but still isn't: The Maze, Standing Rocks, Ernie's Country, The Fins. On the southeast corner of the park there is also a small extension proposed to include beautiful Lavender Canyon.

If you have made the necessary transition from the conventional opinion of what makes beautiful land, and can *see* this country for itself, it will seem exceptionally beautiful to you. All of it is still fairly remote from the inhuman noise, stench and crowding of cities. Its prime quality is wildness, which will disappear if the country is made accessible to windshield tourists. There is, we submit, already more than enough chrome-plated scenery in Utah to satisfy the needs and wants of those who cannot—or will not—leave their cars behind.—*Philip Hyde/Taylorsville, California.*

# The Situation

Since the first edition of *Slickrock* appeared in 1971, Canyonlands National Park has been enlarged to include the Maze; Capitol Reef National Monument has become a National Park with enlarged boundaries; and Glen Canyon National Recreation Area was created by Congress.

As a result of the intensive lobbying by conservationists, the Glen Canyon National Recreation Area was enlarged to include about four-fifths of the proposed Escalante Wilderness. Provisions of this bill include two-year studies of potential wilderness and of "roads necessary for full utilization of the area." Unfortunately, it also authorizes the trans-Escalante highway. The Utah Congressman responsible for this section of the bill was defeated for re-election a month after its passage.

The future of the trans-Escalante highway, as well as the trans-Colorado-Orange Cliffs road, thus depends on the findings of the National Park Service study. The final decisions on these and other roads, as well as specific wilderness-area designations, will be made by Congress sometime after the results of the study are presented in October 1974. While it is quite likely that the proposal for the new trans-Escalante highway will die a natural death—because it is illogical, irresponsible and atrociously expensive—further effort is necessary to insure its demise.

Secretary Rogers Morton has denied permits for the huge Kaiparowits coal-burning, power-generating, industrial complex, but Utah politicians are pressuring him to reverse that decision. Meanwhile, two Utah power companies forge ahead with plans and field studies for powerplants in the Escalante area and damsites on the Escalante River.

Much has been done to protect these wild, beautiful, lonely places. Much remains to be done. We can win many successive steps of a conservation battle, but we can lose only once.

—*June Viavant*
*The Escalante Wilderness Committee*
*Salt Lake City, Utah*

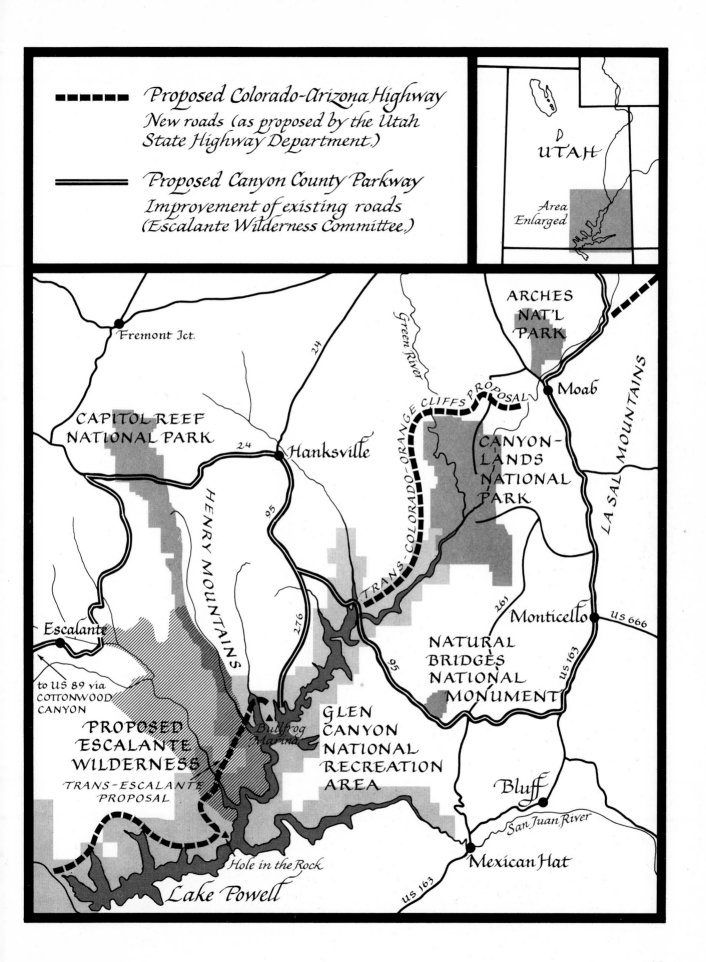

Abbey's part

# I. How it was

The first time I ever had a glimpse of the canyon country was in the summer of 1944. I was a punk kid then, scared and skinny, hitch-hiking around the United States. At Needles, California, bound home for Pennsylvania, I stood all day by the side of the highway, thumb out. Nobody stopped. In fact, what with the war and gaso-line rationing, almost nobody drove by. Squatting in the shade of a tree, I stared across the river at the porphyritic peaks of Arizona, crazy ruins of volcanic rock floating on heat waves. Purple crags, lavender cliffs, long blue slopes of cholla and agave—I had never before even dreamed of such things.

In the evening an old black man with white whiskers crept out of the brush and bummed enough money from me for his supper. Then he showed me how to climb aboard an open boxcar when a long freight train pulled slowly out of the yards, rumbling through the twilight, eastward bound. For half the night we climbed the long grade into Arizona. At Flagstaff, half frozen, I crawled off the train and into town looking for warmth and hospitality. I was locked up for vagrancy, kicked out of jail the next morning, and ordered to stay away from the Santa Fe Railroad. And no hitch-hiking, neither. And don't never come back.

Humbly I walked to the city limits and a step or two beyond, held out my thumb and waited. Nobody came. A little after lunch I hopped another freight, all by myself this time, and made myself at home in a big comfortable empty side-door Pullman, with the doors open on the north. I found myself on a friendly train, in no hurry for anywhere, which stopped at every yard along the line to let more important trains roar past. At Holbrook the brakemen showed me where to fill my canteen and gave me time to buy a couple of sandwiches before we moved on.

From Flag to Winslow to Holbrook; and then through strange, sad, desolate little places called Adamana and Navajo and Cham-bers and Sanders and Houck and Lupton—all the way to Gallup, which we reached at sundown. I left that train when two rough-looking customers came aboard my boxcar; one of them began paring his fingernails with a switchblade knife while the other stared at me with somber interest. I had forty dollars hidden in my shoe. Not to mention other treasures. I slipped out of there quick. Suddenly homesick I went the rest of the way by bus, nonstop, about 2,500 miles, the ideal ordeal of travel, second only to a seasick troopship.

But—I had seen the southern fringe of the canyon country. And did not forget it. For the next three years, through all the misery and tedium, humiliation, brutality and ugliness of my share of the war and the military, I kept bright in my remembrance, as the very picture of things which are free, decent, sane, clean and true, what I had seen—and felt—yes, and even smelled—on that one blazing afternoon on a freight train rolling across northern Arizona.

I mean the hot dry wind. The odor of sagebrush and juniper, of sand and black baking lava-rock. I mean I remembered the sight of a Navajo hogan under a bluff, red dust, a lonesome horse browsing far away down an empty wash, a windmill and water tank at the hub of cattle trails radiating toward a dozen different points on the horizon, and the sweet green of willow, tamarisk and cottonwood trees in a stony canyon. There was a glimpse of the Painted Desert. For what seemed like hours I could see the Hopi Buttes, far on the north, turning slowly on the horizon as my train progressed across the vast plateau. There were holy mountains in the far distance. I saw gleaming meanders of the Little Colorado and the red sandstone cliffs of Manuelito. Too much. And hard-edge cumulus clouds drifting in fleets through the dark blue sea of the sky. And most of all the radiance of that high desert sunlight, which first stuns then exhilarates your senses, your mind, your soul.

But this was only, as I said, the fringe. In 1947 I returned to the Southwest and began to make my first timid, tentative explorations toward the center of that beautiful blank space on the maps. From my base at the University of New Mexico, where I would be trying, more or less, for the next ten years, off and on, to win a degree, I drove my old Chevvie through mud and snow, brush and sand, to such places as Cabezon on the Rio Puerco and from there south to Highway 66. They said there was no road. They were right. But we did it anyhow, me and a kid named Alan Odendahl (a brilliant economist since devoured by the insurance industry), freezing at night in our kapok sleeping bags and eating tinned tuna for breakfast, lunch and supper. Tire chains and skinned knuckles; shovels and blisters; chopping brush to fill in a bog-hole, I missed once and left the bite of the axe blade in the toe of my brand-new Redwing engineer boots. (In those days philosophy students wore boots; now —more true to the trade—they wear sandals, as Diogenes advised, or go barefoot like Socrates.) Next we made it to Chaco Canyon, where we looked amazed at Pueblo Bonito in January. And then to the south rim of Canyon de Chelly—getting closer—and down the foot-trail to White House Ruin. An idyllic place, it seemed then; remote as Alice Springs and far more beautiful. Everett Ruess had wandered here (see Author's Notes).

On one long holiday weekend another friend and I drove my old piece of iron with its leaky gas tank and leaky radiator northwest around the Four Corners to Blanding, Utah, and the very end of the pavement. From there we went by dusty washboard road to Bluff on the San Juan and thought we were getting pretty near the end of the known world. Following a narrow truck road through more or less ordinary desert we climbed a notch in Comb Ridge and looked down and out from there into something else. Out *over* something else. A landscape which I had not only never seen before but which did not resemble anything I had seen before.

I hesitate, even now, to call that scene beautiful. To most Americans, to most Europeans, natural beauty means the sylvan—pastoral and green, something productive and pleasant and fruitful—pastures with tame cows, a flowing stream with trout, a cottage or cabin, a field of corn, a bit of forest, in the background a nice snow-capped mountain range. At a comfortable distance. But from Comb Ridge you don't see anything like that. What you see from Comb Ridge is mostly red rock, warped and folded and corroded and eroded in various ways, all eccentric, with a number of maroon buttes, purple mesas, blue plateaus and gray dome-shaped mountains in the far-off west. Except for the thin track of the road, switchbacking down into the wash a thousand feet below our lookout point, and from there climbing up the other side and disappearing over a huge red blister on the earth's surface, we could see no sign of human life. Nor any sign of any kind of life, except a few acid-green cottonwoods in the canyon below. In the silence and the heat and the glare we gazed upon a seared wasteland, a sinister and savage desolation. And found it infinitely fascinating.

We stared for a long time at the primitive little road tapering off into the nothingness of the southwest, toward fabled names on the map—Mexican Hat, Monument Valley, Navajo Mountain—and longed to follow. But we didn't. We told ourselves that we couldn't: that the old Chev would never make it, that we didn't have enough water or food or spare parts, that the radiator would rupture, the gas tank split, the retreads unravel, the water pump fail, the wheels sink in the sand—fifty good reasons—long before we ever reached civilization on the other side. Which at that time would have been about Cameron, maybe, on U.S. 89. So we turned around and slunk back to Albuquerque the way we'd come, via the pavement through Monticello, Cortez and Farmington. Just like common tourists.

Later, though, I acquired a pickup truck—first of a series—and became much bolder. Almost every weekend or whenever there was enough money for gas we took off, all over New Mexico, over into Arizona, up into Colorado, and eventually, inevitably, back toward the Four Corners and beyond—toward whatever lay back of that beyond.

The words seem too romantic now, now that I have seen what men and heavy equipment can do to even the most angular and singular of earthly landscapes. But they suit our mood of that time. We were desert mystics, my few friends and I, the kind who read maps as others read their holy books. I once sat on the rim of a mesa above the Rio Grande for three days and nights, trying to have a

Cactus and rock. Rock and
cactus. Life anyhow crawls into
light on the tough frontiers of
aridity. Old hedgehog, old
prickly pear. They never give up.

vision. I got hungry and saw God in the form of a beef pie. There
were other rewards. Anything small and insignificant on the map
drew us with irresistible magnetism. Especially if it had a name
like Dead Horse Point, or Wolf Hole, or Recapture Canyon, or Black
Box, or Old Paria (abandoned), or Hole-in-the-Rock, or Paradox, or
Cahone (Pinto Bean Capital of the World), or Mollie's Nipple, or
Dirty Devil, or Pucker Pass, or Pete's Mesa. Or Dandy Crossing.

Why Dandy Crossing? Obvious: because it was a dandy place to
cross the river. So, one day we loaded the tow chain and the spare
spare, the water cans and gas cans, the bedrolls and bacon and
beans and boots into the back of the truck and bolted off. For the
unknown. Well, unknown to us.

Discovered that, also unbeknownst to us, the pavement had been
surreptitiously extended from Monticello down to Blanding, while
we weren't looking, some twenty miles of irrelevant tar and gravel.
A trifling matter? Perhaps. But I felt even then (seventeen years ago)
a small shudder of alarm. Something alien was moving in, something queer and out of place here.

At Blanding we left the pavement and turned west on a dirt road
into the sweet wilderness. Wilderness? It seemed like wilderness to

us. Till we reached the town of Green River 180 miles beyond, we would not see another telephone pole. Behind us now was the last drugstore, the final power line, the ultimate policeman, the end of all asphalt, the very tip of the monster's tentacle.

We drove through several miles of pigmy forest—pinyon pine and juniper—and down into Cottonwood Wash, past Zeke's Hole and onward to the crest of Comb Ridge. Again we stopped to survey the scene. But no turning back this time. While two of my friends walked down the steep and twisting road to remove rocks and fill in holes, I followed with the pickup in compound low, riding the brake pedal. Cliff on one side, the usual thousand-foot drop on the other. I held the wheel firmly in both hands and stared out the window at my side, admiring the awesome scenery. My wife watched the road.

The valley of Comb Wash looked like a form of paradise to me. There was a little stream running through the bright sand, a grove of cottonwoods, patches of grass, the color-banded cliffs on either side, the woods above—and not a house, not a human soul in sight, not even a cow or horse. Eden at the dawn of creation. What joy it was to know that such places still existed, waiting for us when the need arose. At that time I did not realize that what looked so open and free was, even then, tied up in cattle-grazing permits, de facto property of the local ranchers.

We ate lunch by the stream, under the cottonwood trees, attended by a few buzzing flies and the songs of canyon wren and pinyon jay. Midsummer: the cattle were presumably all up in the mountains now, fattening on larkspur and lupine and purple penstemon. God bless them—the flowers, I mean. The wine passed back and forth among the four of us, the birds called now and then, the thin clear stream gurgled over the pebbles, bound for the San Juan River (which it would not reach, of course; sand and evaporation would see to that). Above our heads an umbrella of living, lucent green sheltered us from the July sun. We enjoyed the shade as much as the wine, the birds and flies and one another.

For another twenty miles or so we drove on through the pinyon-juniper woods, across the high mesa south of the Abajo Mountains. The road was rough, full of ruts and rocks and potholes, and we had to stop a few times, get out the shovel and do a little road-work, but this was more a pleasure than otherwise. Each such stop gave all hands a chance to stretch, breathe deep, ramble, look—and see. Why hurry? It made no difference to us where nightfall might catch us. We were ready and willing to make camp anywhere. And in this splendid country, still untouched by development and industrialism, almost any spot would have made a good campsite.

Storm clouds overhead? Good. What's July in the desert without a cloudburst? My old truck creaked and rattled on. Bouncing too fast down into a deep wash I hit a pointed rock imbedded in the road and punched a hole through one of the tires. We installed one of the spares and rumbled on.

Late in the afternoon we reached Natural Bridges. We drove down a steep, narrow, winding dirt track among the pinyon pines—fragrant with oozing gum—and into the little campground. One other car was already there. In other words, the place was badly overcrowded. We debated among ourselves whether to stay or go on. The girls wanted to stay, so we did. We spent the next day in a leisurely triangular walk among the three great bridges—Owachomo, Sipapu and Kachina—and a second night at the little Park Service campground. It was the kind of campground known as "primitive," meaning no asphalt driveways, no flush toilets, no electric lights, no numbered campsites, no cement tables, no police patrol, no fire alarms, no traffic controls, no movies, slide shows or press-a-button automatic tape-recorded natural history lectures. A terrible, grim, deprived kind of campground, some might think. Nothing but stillness, stasis and stars all night long.

In the morning we went on, deeper into the back country, back of beyond. The "improved" road ended at Natural Bridges; from here to the river, forty-five miles, and from there to Hanksville, about another forty, it would be "unimproved." Good. The more unimproved the better, that's what we thought. Assuming, of course, in those innocent days, that anything good would be allowed to remain that way.

Our little road wound off to the west, following a sort of big bench, with the sheer cliffs of a plateau on the south and the deep, complicated drainages of White Canyon on the north. Beyond White Canyon were Woodenshoe Butte, the Bear's Ears, Elk Ridge and more fine blank areas on the maps. Nearby were tawny grass and buff-colored cliffs, dark-green junipers and sandstone scarps.

As we descended gradually toward the river the country opened up, wide and wild, with nowhere any sign of man but the dirt trail road before us. We liked that. Why? (*Why* is always a good question.) Why not? (Always a good answer.) But why? One must attempt to answer the question—someone always raises it, accusing us of "disliking people."

Well then, it's not from simple misanthropy. Speaking generally, for myself, I like people. Though not very much. Speaking particularly, I like some people, dislike others. Like everyone else who hasn't been reduced to moronism by our commercial boy scout ethic, I like my friends, dislike my enemies, and regard strangers with a tolerant indifference. But why, the questioner insists, why do people like you pretend to love uninhabited country so much? Why this cult of wilderness? Why the surly hatred of progress and development, the churlish resistance to all popular improvements?

Very well, a fair question, but it's been asked and answered a thousand times already, enough books to drive a man stark naked mad have dealt in detail with the question. There are many answers, all good, each sufficient. Peace is often mentioned; beauty; spiritual refreshment, whatever that means; re-creation for the soul, whatever that is; escape; novelty, the delight of something different; truth and understanding and wisdom—commendable virtues in any man, anytime; ecology and all that, meaning the salvation of variety, diversity, possibility and potentiality, the preservation of the genetic reservoir, the answers to questions which we have not yet even learned to ask, a connection to the origin of things, an opening into the future, a source of sanity for the present—all true, all wonderful, all more than enough to answer such a dumb dead degrading question as "Why wilderness?"

To which, nevertheless, I shall append one further answer anyway: *because we like the taste of freedom. Because we like the smell of danger.*

Descending toward the river the junipers become scarce, give way to scrubby, bristling little vegetables like black brush, snakeweed and prickly pear. The bunch grass fades away, the cliff rose and yucca fall behind. We topped out on a small rise and there ahead lay the red wasteland again—red dust, red sand, the dark smoldering purple reds of ancient rocks, Chinle, Shinarump and Moenkopi, the old Triassic formations full of radium, dinosaurs, petrified wood, arsenic and selenium, fatal evil monstrous things, beautiful, beautiful. Miles of it, leagues of it, glittering under the radiant light, swimming beneath waves of heat, a great vast aching vacancy of pure space, waiting. Waiting for what? Why, waiting for us.

Beyond the red desert was the shadowy crevasse where the river ran, the living heart of the canyonlands, the red Colorado. Note my use of the past tense here. That crevasse was Glen Canyon. On either side of the canyon we saw the humps and hummocks of Navajo sandstone, pale yellow, and beyond that, vivid in the morning light, rich in detail and blue in profile, the Henry Mountains, last-discovered (or at least the last-named) mountain range within the coterminous United States. These mountains were noted, as one might expect, by Major John Wesley Powell, and by him named in honor of his contemporary, Joseph Henry, secretary of the Smithsonian Institution. Beyond the mountains we could see the high Thousand Lake and Aquarius plateaus, some fifty miles away by line of sight. In those days, of course (seventeen years ago!), before the potash mills, cement plants, uranium mills and power plants, fifty miles of clear air was nothing—to see mountains 100 miles away was considered commonplace, a standard of vision.

We dropped down into that red desert. In low gear. Moved cautiously across a little wooden bridge that looked like it might have been built by old Cass Hite himself, or even Padre Escalante, centuries before. Old yellow-pine beams full of cracks and scorpions, coated with the auburn dust. Beneath the bridge ran a slit in the sandstone, a slit about ten feet wide and 100 feet deep, so dark down in there we could hardly make out the bottom. We paused for a while to drop rocks. The sunlight was dazzling, the heat terrific, the arid air exhilarating.

I added water to the radiator, which leaked a little, like all my radiators did in my student days, and pumped up one of the tires, which had a slow leak, also to be expected, and checked the gas tank, which was a new one and did not leak, yet, although I could see dents where some rocks had got to it. We all climbed aboard and went on. Mighty cumuli-nimbi massed overhead—battleships of vapor, loaded with lightning. We cared not a fig.

Coyote Gulch in the Escalante wilderness. The developers want to build a highway through this region to make it "accessible." But how did Phil Hyde get in here to take a picture? Why the damn fool walked, I suppose, or maybe floated in on angel wings.

We jounced along in my overloaded pickup, picking our way at two miles an hour in and out of the little ditches—deadly axle busters—which ran across, not beside the road, heading the side canyons, climbing the benches, bulling our way through the sand of the washes. We were down in the land of standing rock, the world of sculptured sandstone, crazy country, a bad dream to any dirt farmer, except for the canyon bottoms not a tree in sight. Good.

We came to the crossing of White Canyon, where I gunned the motor hard, geared down into low, and charged through the deep sand. Old cottonwoods with elephantine trunks and sweet green trembling leaves caught my eye. Lovely things, I thought, as we crashed over a drop-off into a little stream. Glimpsed sandpipers or killdeer scampering out of the way as a splash of muddy water drenched the windshield. Hang on, I said. Heard a yelp as my friend's girl friend fell off the back of the truck. Couldn't be helped. The truck lurched up the farther bank, streaming with water, and came to halt on the level road above, more or less by its own volition.

We all got out to investigate. Nobody hurt. We ate our lunch beneath the shade of the trees. In the desert, under the summer sun, shade makes the difference between intolerable heat and a paradisiacal coolness. The temperature drops thirty degrees inside the shadow line, if there is free ventilation. One of the interesting things this means is that if homes and public buildings in the

Southwest were properly designed, built for human pleasure instead of private profit, there would be no need for artificial air conditioning. Is that asking too much from those who design our lives? Yes it is. The humblest Papago peasant or Navajo sheepherder knows more about efficient hot-country architecture than a whole skyscraper full of Del Webbs or R. Buckminster Fullers.

Why contaminate a reminiscence of something clean and sane and fit for human consumption with any reference to the squalid rottenness of, for example, Phoenix, Arizona? Would it not be better to remember the Southwest as it was—and write it off? You have to do it sooner or later. Do it now.

I won't.

We went. (Black shadows on the rosy rock. One buzzard, red-

Out of the shade and into the heat: a cattle drive near Fruita, Capitol Reef. A hard way to make a living—and a good one.

necked bald-headed black-winged anarchist, hangs in the blue, floats on the thermal, regarding us with a tolerant indifference.)

After the siesta, in midafternoon, we drove up from the ford and around a bench of naked rock several miles long and through a notch or dugway in a red wall. Below us lay Hite, Dandy Crossing, the river.

We descended, passed a spring and more cottonwoods, and came to the combination store, gas station and post office which was not only the business center but almost the whole of Hite. At that time I believe there were no more than three families living in the place, which must have been one of the most remote and isolated settlements in the forty-eight states. There were also a few miscellaneous individuals—prospectors, miners, bums, exiles, remittance men— hanging about. The total population fluctuated from year to year with the fortunes of the uranium industry. Eventually the dam was built, the river backed up, and everybody flooded out.

We stopped to buy gas—fifty cents a gallon, cheap at the price— and a round of beer. I met Mr. Woody Edgell, proprietor, who was already unhappy about the future prospects of Glen Canyon. He took a dim view both of the dam and of the Utah State Highway Department's proposed bridge-building schemes for the vicinity. Not because they would put him out of business—they wouldn't; he could relocate—but because he liked Hite and Glen Canyon the way they were, neolithic.

Not everybody felt that way. I talked with a miner's wife and she said that she hated the place, claimed that her husband did too, and said that only lack of money kept them there. She looked forward with gratitude to the flooding of Hite—a hundred feet under water was not deep enough, she thought. She'd be glad to be forced to leave. Well, that's the way most women *would* feel about a place like Hite; it was not the kind of place that women like.

There was a middle-aged fellow sitting outside the store, on a bench in the shade, drinking beer. He had about a month's growth of whiskers on what passed for a face. I bought him another can of Coors and tried to draw him into conversation. He was taciturn. When I asked him what he did around there he looked up at the clouds and over at the river and down at the ground between his boots, thinking hard, and finally said: "Nothing."

A good and sufficient answer. Taking the hint, I went away from there, leaving him in peace. My own ambition, my deepest and truest ambition, is to find within myself someday, somehow, the ability to do likewise, to do nothing—and find it enough.

Somewhat later, half waterlogged with watery beer, we went for a swim in the river, stark naked, and spouted silty water at the sky. The river tugged at our bodies with a gentle but insistent urge:

*Come with me,* the river said, *close your eyes and quiet your limbs and float with me into the wonder and mystery of the canyons, see the unknown and the little known, look upon the stone gods face to face, see Medusa, drink my waters, hear my song, feel my power, come along and drift with me toward the distant, ultimate and legendary sea....*

Sweet and subtle song. Perhaps I should have surrendered. I almost did. But didn't. We piled ourselves wet and cooled and strangely tired back into the old truck and drove on down the shore to the ferry crossing.

This was perhaps a mile beyond the store. There was a dirt-covered rock landing built out from shore, not far, and a pair of heavy cables strung across the river to the western bank. The ferry itself was on the far side where Art Chaffin, the ferryman, lived in a big house concealed by cottonwoods. A man appeared among the trees on the opposite shore, stepped aboard his ferry and started the engine, engaged the winch. The strange craft moved across the river's flow toward us, pulling itself along the sagging cable. It was not a boat. It appeared to be a homemade barge, a handmade contraption of wood and steel and baling wire: gasoline engine, passenger platform, vehicle ramp and railings mounted on a steel pontoon. Whatever it was, it worked, came snug against the landing. I drove my pickup aboard, we all shook hands with old Art Chaffin and off we went, across the golden Colorado toward that undiscovered West on the other side.

The Hite Ferry had a history, short but rich. Following old Indian trails, Cass Hite came to and named Dandy Crossing in 1883. It was one of the very few possible fords of the river in all the 240 miles or so between Moab and Lee's Ferry. That is, it could be negotiated by team and wagon during low water (late summer, winter). But it did not become a motor vehicle crossing until 1946, when Chaffin built his ferry. The first ferry sank in 1947; Chaffin built a second, which he sold in 1956 to a man named Reed Maxfield. In 1957 Reed Maxfield had an accident and drowned in the river. His widow kept the ferry in operation until a storm in November of 1957 tore the barge loose from its mooring and sank it. By this time the ferry had become fairly well known and its service was in some demand; the State Highway Department was obliged to rebuild it. Mrs. Maxfield was hired to continue running it, which she did until Woody Edgell took over in 1959. He was the last ferryman, being finally flooded out by the impounded waters of Glen Canyon Dam in June, 1964. To replace the ferry the Utah Highway Department had to build not one but three bridges: one over the mouth of the Dirty Devil, one over the Colorado at Narrow Canyon, and the third over White Canyon. Because of the character of the terrain in there—hard to believe unless you see it for yourself—there is no other feasible way to get automobiles across the canyons. Thus, three big bridges, built at the cost of several million dollars (your money and mine, not the Utah Highway Department's), were required to perform the same service which Art Chaffin's home-designed ferry had provided quite adequately, and much more beautifully, for eighteen years. This being only an infinitesimal fraction of the total price we have all had to pay for that grand monument to stupidity known as Glen Canyon Dam. (Let all the Faithful living downriver from this dam take warning: there are thousands of us who never forget, during our bedtime prayers, to ask God for one little precision earthquake in the immediate vicinity of Glen Canyon

Dam. We are devout; we have faith; and someday soon our prayers shall be answered.)

Back to 1953: As we were leaving the river, Mr. Chaffin, glancing at the clouded sky, advised me to watch for flash floods in North Wash.

North Wash? I said. Where's that?

Where you're going, he said.

So it was.

The road to Hanksville and from there to Green River led through North Wash. The only road.

We followed the right bank of the river for a couple of miles upstream, rough red cliffs above shutting off the view of the mountains and high country beyond. The sky was dark all right. The willows on the banks were lashing back and forth under a brisk wind and a few raindrops exploded against the windshield.

Somebody suggested camping for the night beside the river. A good idea. But there was one idiot in our party, I won't identify him, who was actually *hoping* to see a flash flood. And he prevailed. In the late afternoon, under a turbulent sky, we turned away from the river and drove into a deep, dark, narrow canyon leading west and north, where the road (you might call it that) wound up and out, toward the open country twenty miles above. According to my Texaco road map. Which also said, quote, *Make local inquiry before attempting travel in this area.*

A fine map. A lovely quote. All maps should be sprinkled with such remarks.

A good canyon. A little creek came down it, meandering casually from wall to wall, in the rhythm of its force. Vertical walls. The road we followed crossed that stream about ten times per mile, out of necessity. I tested the brakes occasionally. Wet drums. No brakes. But it hardly mattered, since we were gradually ascending. The little sprinkling of rain had stopped and everyone admired the towering canyon walls, the alcoves and grottoes, the mighty boulders big as boxcars strewn about on the canyon floor. The air was cool and sweet, the tamarisk and redbud and box elders shivered in the breeze on their high alluvial benches. Flowers bloomed, as I recall. Birds chirruped now and then, humble and discreet.

I became aware of a deflating tire and stopped the pickup right in the middle of the wash, spanning the little rivulet of clear water. It was the only level place immediately available.

The two girls walked on ahead up the road while my buddy and I jacked up the truck and pulled the wheel. We checked the tire and found that we'd picked up a nail somewhere, probably down by the river.

We were standing there bemused and barefooted, in the stream, when we heard the women begin to holler from somewhere out of sight up the canyon. Against the background noise of the constant wind, and something like a distant waterfall, it was hard to make out their words exactly.

Mud? Blood? Crud?

As we stood there discussing the matter I felt a little sudden surge in the flow of water between my ankles. Looking down, I saw that the clear water had turned into a thick, reddish liquid, like tomato soup. Like blood. Like tired old venous blood pouring from a dirty wound.

Our spare spare was packed away beneath a load of duffle, pots and pans and grub boxes. So we jammed the flat tire back on and lugged it down quick with a couple of nuts. I could hear both girls running toward us, still hollering. My friend picked up the hub cap before it floated away with the rest of the wheel nuts, and stared up the canyon. We couldn't see anything much yet but we could hear it—a freight train rolling full speed down North Wash.

We jumped in the truck, I started the motor and tried to drive away. The engine roared vigorously but nothing else moved. One wheel still jacked off the ground. No positive traction in that old pickup. We had to get out again and then discovered that the jack handle had disappeared under the soup. We pushed the truck forward, off the jack, and discovered that it was still in gear. The truck humped ahead over some rocks and stalled. The main body of the flood now appeared around the bend up canyon. Magnificent spectacle. We got back in the truck, got the motor going, and lurched and yawed, flat tire flopping, out of the bottom of the wash and up onto the safety of higher ground. As the flood roared past below.

The girls joined us. There was no rain at all where we were, and the ground was perfectly dry. But you could feel it tremble with the resonance of the flood. From within the flood, under the rolling red waters, you could hear the grating and grumble of big rocks, boulders, as they clashed on one another, a sound like the grinding of molars in a pair of leviathan jaws. The kind of sound, in other words, for which neither imagination nor fantasy can ever really prepare you. The unbelievable reality of the real.

Our road was cut off ahead and behind. We camped on the bench that evening, made supper in the dense violet twilight of the canyon while thousands of cubic tons of semiliquid sand, silt, mud, rock, uprooted junipers, logs, a dead cow, rumbled by about twenty feet away. Deep and rich as our delight.

The juniper fire smelled good. The bacon and beans were better. A few clear stars switched on in that narrow slot of sky between the canyon walls overhead. We built up the fire and sang. My wife was beautiful. My friend's girl friend was beautiful. My old pickup truck was beautiful.

Sometime during the night the flood dropped off and melted away, almost as abruptly as it had come. We awoke in the morning to the music of canyon wrens and a trickling stream, and found that our road was still in the canyon, same canyon, though kind of folded over and tucked in and rolled up in corners here and there. It took us considerable roadwork, much hard labor and all day long to get out of North Wash. And it was worth every minute of it. Never had such interesting work again till the day I tried to take a Hertz rental Super-Sport past Squaw Spring and up Elephant Hill in The Needles. Or the time another friend and I carried his VW

Capitol Reef, old road, clay hills in foreground, clouds to watch. Watch those clouds: they have much to say.

Beetle down through Pucker Pass off Dead Horse Point after a good rain. Or the time—well, it's a long list.

At North Wash we had a nice midday rest at Hog Spring, halfway out. We met a prospector in a jeep coming in. He said we'd never make it. Hogwash. We said he'd never make it. He looked as pleased as we were, and went on.

Today the old North Wash trail road is partly submerged, the rest obliterated. The state has ripped and blasted and laid an asphalt highway through and around the area to link the fancy new tin bridges with the outside world. The river is gone, the ferry is gone, Dandy Crossing is gone. Most of the formerly primitive road from Blanding west has now been improved beyond recognition. All of this, the engineers and politicians and bankers will tell you, makes the region now easily accessible to everybody, no matter how fat, feeble or flaccid. That is a lie.

It is a lie. For those who go there now, smooth, comfortable, quick and easy, gliding through as slick as grease, will never be able to see what we saw. They will never feel what we felt. They will never learn what we know.

31

# 2. A history: natural and otherwise

The canyon, mesa and slickrock country, as we shall define it in this book, lies almost entirely within southeast Utah. On the north it is bounded by the Tavaputs Plateau, the south-facing escarpment of which is called locally the Roan Cliffs, from the color, or the Book Cliffs, from the stratigraphy. Winding in an east-west direction for 215 miles from Price, Utah, almost to Grand Junction, Colorado, the Book Cliffs present what Major Powell called "one of the most wonderful façades in the world."

The east boundary of our territory has no such clear definition. We might call it the edge of the Great Sage Plain, that generally rolling tableland which extends from the La Sal Mountains (near Moab) on the north down to the Abajo or Blue Mountains (near Monticello and Blanding) on the south. From there you could use the Comb Ridge monocline to its juncture with the San Juan River as the remainder of the eastern boundary.

For a southern boundary we shall designate the San Juan River to its confluence with the Colorado River near Navajo Mountain, and from there the augmented Colorado to the mouth of Glen Canyon at Lee's Ferry in the extreme north of Arizona. South of these rivers is Navajo country; beyond Lee's Ferry begins that part of Grand Canyon known as Marble Gorge and then the Grand itself—*The* Canyon—and all that is something else. Much of the Navajo Reservation, too, lies within what geographers call the Colorado Plateau Province, and shares the major characteristics of the canyon country but like Grand Canyon it is a different world, a world in itself, far too much to attempt to include here.

As the western boundary, we begin at Lee's Ferry and go northwest to the headwaters of the Paria River near Bryce Canyon National Park. From there, along a line running approximately north-northeast we follow the base of the great plateaus of central Utah— the Paunsaugunt, Table Cliff, Aquarius (also known as Boulder Mountain), Thousand Lake and Wasatch—back full circle to Price, Castle Gate and the western end of the Book Cliffs.

Enough. Within this underslung lopsided rump-sprung dough-bellied highly irregular parallelogram lies the least inhabited, least inhibited, least developed, least improved, least civilized, least governed, least priest-ridden, most arid, most hostile, most lonesome, most grim bleak barren desolate and savage quarter of the state of Utah—the best part by far. So far. (The problem now is to keep the growth-and-progress swine from getting their fatal grip on it.) There is enough interesting country here—most of it perpendicular to most of the rest—to keep a man happy, healthy and high through an entire series of carnations, incarnations and reincarnations.

What do we have within these rough boundaries? Think of it as being essentially a great desert plateau, averaging a mile above sea level. Rising above the general surface are a few isolated mountains, islands of cloud and forest surrounded by a waste of largely naked sandstone: the slickrock wilderness. Sinking below the general surface, in some places two or three thousand feet below it, are the canyons. Hundreds of canyons—major, secondary, minor—all finally converging, like the twigs, branches and limbs of a tree, upon the central canyon of the master stream, the Colorado River.

The island mountains are the Henrys, the Abajos and the La Sals, their high points ranging from 9,500 feet to nearly 13,000. All are of the type called laccolithic, meaning of volcanic origin. But they are not and never were true volcanoes; according to current and prevailing geomorphology, the mountains were formed by intrusions of very hard, very resistant rock—dioritic porphyry—forced up from the mantle into, but not through, the crust of the earth. As the intrusions rose they carried the surface rock up with them; eventually the surface rock was worn away by erosion, revealing to the speculative eyes of geologists (such men as Powell, Dutton, Gilbert, Gregory, Hunt and Baker among the principals) the more durable and more significant rock beneath.

Despite their respectable height, forest cover and often snow-covered peaks, these desert mountains do not attract enough precipitation or produce enough drainage to explain the surrounding maze of canyons, the remnant buttes and mesas, the erosional rock jungles. These were and are primarily the result of the work of the Green, the Colorado and the San Juan rivers whose headwaters lie far away in the mountains of Wyoming and Colorado. The unusual combination of major rivers carving a way to the sea, at a steep gradient, through an extremely arid plateau, gives us the key to an understanding of the origin of the spectacular landscape of southeast Utah. Without the major rivers we would have had a desert of the central Australian character; without the aridity we would have had river valleys more like those of the Missouri; without the gradient there could have been no main canyon and no elaborate system of tributary canyons.

The rocks of the canyon country are chiefly sandstone in its many variations, with a few formations of shales and mudstones. The oldest exposed formations are those of the Pennsylvanian period, best seen at the goosenecks of the San Juan River near Mexican Hat. Above those, on view, for example, around the western and northwestern base of the Henry Mountains, are the Permian shales. Rest-ing on the Permian formations are the Triassic red beds—Chinle, Shinarump and Moenkopi rock (uranium, petrified wood)—and above them the great cliffs, domes, fins, arches and bridges of the Jurassic sandstones—Wingate, Kayenta, Navajo, Entrada—which form the most abundant and most striking features of the canyonlands. Other dramatic rock forms occur in the San Rafael group of sandstones, particularly in the Cedar Mesa formations (White Rim, The Maze, The Needles). Scattered about on top of all these sandstone deposits are the shales (also containing uranium) and clays of the Morrison and Dakota formations. There are no volcanoes, cinder cones or lava beds within the bounds of the region I am sketching here. All of the rocks mentioned above are millions of years old, of course, if we accept the basic principles of contemporary geology and paleontology, and most are the sedimentary deposits of an ancient inland sea. One exception to this latter case is the aeolian sandstone—petrified, cross-bedded dunes—of the Navajo formation (Arches, Dry Valley, Glen Canyon, Zion).

Penstemon in Coyote Gulch. Talent springs up everywhere. What we need to complete this picture is an antique sardonic rusty sardine can.

Other unusual and prominent aspects of the region are the great monoclines such as Comb Ridge and the Waterpocket Fold, the curious salt sinks or grabens in The Needles and at Arches, the vast up-warp of the San Rafael Swell with its surrounding reef of sandstone fangs, and the geological oddity called Upheaval Dome (a syncline) in Canyonlands National Park. Upheaval Dome collapsed long ago, apparently because of the leaching away of huge salt deposits underneath, so that what you see if you go there is actually a crater full of rock ruins and the fragmentary remains of the dome on the rim of the crater.

We also have a few arsenic springs, sulfur springs, hot springs and geysers.

The principal canyons are those of the three rivers. The most beautiful, intricate, and important of all, the very living heart of the canyonlands, was Glen Canyon, at present submerged beneath the waters of a dam built near Page, Arizona. Of the canyons that remain, the most extensive and the one which most resembles Glen Canyon is the canyon system of the Escalante. Of equal beauty and value (in my opinion) are the canyons of the Green River—Labyrinth and Stillwater—that fall within the province of this book. Deepest and most rugged is Cataract Canyon, which drops 425 feet within a distance of forty miles, or more than ten feet to the mile; there are more than forty rapids here, some of them as wild as any in the Grand Canyon. The canyon of the Colorado between Moab and the river's confluence with the Green is also of great beauty and interest, passing through dramatic walls of sandstone which much resemble the former Glen Canyon. As for the canyons of the San Juan River I can say little about them, having seen only what is apparent from Muley Point and the Goosenecks Overlook, both in the vicinity of Mexican Hat.

Many of the tributary canyons come close to equaling the river canyons in magnitude and dramatic form. The Escalante I have already mentioned, and will describe in more detail later. The many deep, narrow, fantastic canyons that formerly led into Glen Canyon (most of them at grade level) are now, like the master canyon, temporarily lost to us, accessible only to scuba divers, but below the dam at Page there remains a fourteen-mile vestige of the great Glen Canyon, still undamaged except for the change in the character of the river caused by the dam. Entering the mouth of Glen Canyon at Lee's Ferry is the Paria (or Pariah), a splendid canyon more than forty miles in length, of which I shall also have more to say.

Other side canyons of great interest are those such as Davis Gulch, Coyote Gulch, Stevens Canyon, Death Canyon and Silver Falls Canyon which lead into the Escalante; Gypsum Canyon and Dark Canyon off the Cataract; Grand Gulch off the San Juan; the many canyons—Horse, Horsethief, Horseshoe and others, some of them not even named—which lead into the Green River; the spectacular canyons that drain The Needles area—Salt, Indian Creek, Lavender, Elephant, and Red; the canyon of the Dirty Devil above the former site of Hite Ferry or Dandy Crossing; the grim, evil, poisonous, sulfur-smelling and little-known canyon called Onion Creek (one of my secret favorites and I'll not reveal its location here); Buckskin Canyon (a crazy place); Courthouse Wash (in the Arches); the marvelous canyons, like Muley Twist, in the Waterpocket Fold; Black Box Canyon (can't seem to recall where this one is either); Cane Springs Canyon, near Moab; Mill Creek, Last Chance and Salvation canyons; the logical branch-work of canyons in The Maze, The Reef and The Swell; and White Canyon; and Happy Canyon; and Cohabitation Canyon; and of course Pucker Pass; and others, many others, all the others, the multitude of happy canyons known and unknown, named and nameless, great and small, beautiful and malignant, grim and golden, each in its way a place of marvels, no two alike (see one and you've seen only one), which help drain and desiccate the general area which we are calling, in this book, the slickrock country.

One way to approach a notion of the pattern of this land is by studying the topographic maps, which are in their own way often works of art. By the use of contour lines for different elevations these maps represent the actual character of the landscape about as closely as a map can—often better than the aerial photographs on which they are based.

The next way to get a broad overview of the country is by flying over it, not too high, in a very slow small airplane. This will give you an adequate general conception of the startling color patterns, the elaboration of the drainage systems (and they are *systems*, despite the original impression of maze-like complexity) and the extensiveness of our subject. But airplanes are noisy, shaky, restless and irritable machines, incapable of hovering in one place, incapable of the kind of silence necessary for contemplation (and contemplation is necessary here), and therefore, except as a most superficial introduction, quite unsatisfactory. Much better would be a sailplane; best of all a pair of feathered wings of your own. Set your heart on becoming, not an angel (angels can't soar), but a hawk, an eagle or—the ultimate—a louse-ridden, carrion-eating, baldheaded turkey vulture. Aim high. Be a buzzard.

Even an automobile is better than an airplane. At least you can stop the damned thing, shut off the engine, get out, step away from the stink of gasoline and oil and actually *see* something, *hear* something, *smell* something, *feel* something. But not much.

Not much. The only right way to get to know this country (any country), the only way, is with your body. On foot. Better yet, on hands and knees. Best of all—after scrambling to a high place—on your rump. Pick out a good spot and just sit there, not moving, for about a year. (This is my own highest ambition.) Keep your eyeballs peeled and just sit there, through the hours, through the days, through the nights, through the seasons—the freeze of winter, the stunning glare and heat of summer, the grace and glory of the spring and fall—and watch what happens.

Where? I can think of several good high places. Somewhere near Grandview Point. Or Temple Mountain in The Swell. Or Land's End. Or the Bears Ears. On the verge of the Kaiparowits Plateau. Woodenshoe Butte. Cleopatra's Chair. Upheaval Dome. Or any one of the

Young hawk on rock in Soda Gulch, slickrock territory. That rock it's brooding on, though very old, contains the egg of the future. Stick around.

island mountains. Pick your place and stay there. You will become a god.

From a site near the Bears Ears, for example, you can gaze out upon the corners of four states. On the east, if the air is clear, you will be able to see the San Juan Mountains and Ute Mountain in southwestern Colorado. To the southeast is the volcanic plug called Shiprock in New Mexico and the Carrizo and Chuska Mountains of extreme northeast Arizona, all within the Navajo reservation. Southward is the Monument Up-warp, with the great buttes of Monument Valley on the skyline. Closer by is Comb Ridge, the Grand Gulch plateau, and the various washes and canyons that meander to the San Juan River. Grand Gulch itself, though no more than a thousand feet deep at any point (no great depth in this territory), is walled in by curved, slick, overhanging bluffs on all sides, difficult of access except by foot. To the southwest you can see the Clay Hills, the Moss Backs, the Red House Cliffs, and far beyond those, the lonely dome of Navajo Mountain, the grand profile of the Kaiparowits Plateau, the misty mountain plateaus beyond. To the west, and below, are the three great bridges of Natural Bridges National Monument in White and Armstrong canyons; beyond are Red Canyon, the remains of Glen Canyon, and the Henry Mountains. Stretching behind this vantage point on the north is long 8,000-foot-high Elk Ridge, which shuts off the view of The Needles and The Confluence. On your northeast are the Abajo Mountains. In all directions you will see not only a spatial panorama but a symphony of color: gray-green mountains, umber plains, maroon canyons with ivory rims, and mesas, buttes, plateaus veiled in various and always changing tones of blue, purple and amber, depending on the time of day, the weather, the quality of the light.

A few paragraphs on the biota:

High on the mountains grow spruce, fir and quaking aspen— usually at elevations above 8,500 feet. Below this zone but inter-graded with it are groves of ponderosa pine and thickets of gambel oak; in moist, cool, shady places you may find Douglas fir.

Pinyon pine and Utah juniper, most characteristic of canyonland flora, grow in the 4,000 to 7,000 foot belt; associated with these beautiful and fragrant little trees, which may occur either in dense stands or as isolated individuals, are such shrubs as cliff rose (pretty as blossoming peach, in fragrance similar to orange trees in bloom), the strange and lovely variety of yucca known as Spanish bayonet, and the big sage, *Artemisia tridentata,* which also bears a sweet and poignant perfume—one of those key and unforgettable odors which have the power to invoke, in themselves, an entire landscape, an entire way of life, a constellation of myths and legends.

At elevations below 4,000 feet, except along water courses, the vegetation becomes sparse: here you will find few perennials but the scrubby shrub called black brush, occasional prickly pear, a type of saltbush, some fishhook and hedgehog cactus.

In canyon bottoms and along the very rare permanent streams the most conspicuous plants are the Fremont poplar, better known as cottonwood, and willow and tamarisk. (The latter, an exotic im-ported from North Africa back in the 1920s, was meant to control erosion along southwestern waterways but immediately got out of control; one more good example of muddleheaded meddling by technologists. Nevertheless it too is an attractive thing.)

Where water is abundant, as around springs and under seeps in canyon walls, still another distinctive plant community can be found, of which the most common members are canes and tules, maidenhair fern, helleborine orchid, monkey flower, bluestem, panic grass and poison ivy.

An immense variety of flowering annuals live here too, at least potentially; their germination, growth and blossoming depend upon both the amount and the spacing of precipitation, a highly erratic factor in this desert climate. When the winter rains and snows come right, April and May will produce a dazzling flower show; if not, there may be almost nothing to see. Much also de-pends upon the chance of locale: one area, well favored, may be carpeted with flowers while another, only a mile away, can be bar-ren. There is also usually a display of flowers in late summer fol-lowing the thunderstorms of July and August but then too, even more than in the spring, much depends upon the luck of the winds and the vapors. In any case, the most common of the flowering an-nuals are regional varieties of globemallow, primrose, verbena, sunflower, aster, datura, penstemon, gilia, princess plume, bee-plant, rock cress, and Indian paintbrush. These are the flowers of the slickrock country, the great rock garden; at higher levels, of course, grow different flowers; in the mountains are lupine, larkspur and the rare blue columbine—among a multitude of others.

Animals of many kinds and small populations call this arid region home. Among the mammals, the biggest and most common is the mule deer. There are some antelope. There may be a few black bear in the La Sal Mountains but this is hardly more than a rumor. Mountain lions are scarce but still around—I have seen their tracks in Escalante, along the Green, and up in the sandy basins on top of the mesas and plateaus. Given the status of big game in the state of Utah, the mountain lion may survive, but nobody really knows how many remain; sadly vulnerable to traps and dogs, this rare, beautiful and splendid beast has at best only a fair chance of making it into the twenty-first century. (About the same as ours, I reckon.) A few bighorn sheep still hold out in the wildest and ruggedest parts of the canyons—I have seen *their* scat on the ledges above the Paria, for example.

I'm not sure how well the coyotes are doing in the canyon country but I know they are not near so common as they should be; the dull-spirited scoundrels who distribute poison all over the landscape for the U.S. Interior Department's Division of Wildlife Services (yes, that's what it's called!) have been at their foul and cowardly work for many years. Still, the net result of their efforts may have been simply to promote the evolution (through unnatural selection) of a subtler and keener and always more intelligent coyote. Although you seldom see or hear them in this region you'll find their neat and tidy signs along most any trail, wash, ledge, canyon floor or other convenient passageway.

Also present, if rarely seen, are gray fox and kit fox, skunk, bobcat and badger, beaver in the rivers, ring-tailed cat and porcupine. Jack rabbits, cottontail rabbits, ground squirrels, pocket gophers and kangaroo mice are nearly everywhere, easily observed. The foolish poisoning and trapping of their natural enemies has inevitably led to a destructive population explosion among these hares, lagomorphs and rodents—the proletariat of the mammalian order.

Rattlesnakes are common, especially the diamondback and a smaller species called horned rattler or faded pygmy. Gopher snakes are present but not so common; also red racers, sand snakes and sidewinder rattlesnakes.

Of the many lizards in these parts, the biggest and most striking are the chuckawallas and the collared lizards. Either will bite if cornered but I am sorry to say that we have no actually poisonous lizard in the canyon country. Presumably the winters are too severe for such an interesting reptile as the Gila monster.

Other crawling creatures you might like to keep in mind, for they are abundant, vigorous, and aggressive in southeast Utah, are scorpions (the breed known formally as giant hairy desert scorpion), centipedes, millipedes, black widow spiders, tarantula spiders, conenosed kissing bugs, and various sundry mites and ticks (some of them disease-bearing, all of them bloodsucking).

Among the birds, the most conspicuous is the great soaring buzzard or turkey vulture. Red-tail and sharp-shinned hawks are com-

Wild lilies on the dune. Their trailing leaves, like a compass, will inscribe perfect arcs and circles in the sand when the wind blows.

mon; golden eagles, though rare, may be seen on occasion. Ravens, magpies, pinyon jays, black-throated sparrows, cliff swallows and mourning doves are abundant. Rock wrens and canyon wrens are common—the call of the latter is surely one of the sweetest and most characteristic sounds of this region. Along the rivers you may see, from time to time, the great blue heron, the snowy egret, the wood ibis and various wild ducks, coots and grebes. Most anywhere you go you will hear and sometimes see the great horned owl. Nighthawks, or "goatsuckers," appear in swirling flocks over the desert during the springtime dawns and evenings—a thrilling sight. Some vireos and warblers nest in the river canyons; bluebirds and thrushes, as one would expect, are common at higher elevations on the mountainsides.

The human history:

The Indians were here first. They discovered America, explored it and settled it, and in so doing did not overlook even the most obscure canyons of the Southwest. Within the area I have delimited as slickrock country you will find abundant traces, messages, relics and ruins of their former occupancy. Although there are no abandoned towns or cliff dwellings here that can match those of Bet-atakin, Keet Seel, Chaco, Hovenweep or Mesa Verde, the Anasazi— "the ancient ones"—did leave a great number of storage huts, stone villages and masonry watchtowers in many sites throughout the canyonlands. Most of these, widely scattered, have had no protection from pothunters or from the vastly more destructive vandalism of the Bureau of Reclamation—some of the finest of the ancient ruins now lie beneath the waters of the Glen Canyon Dam impoundment. Most of them have no protection now, except on paper, and therefore I think it wiser not to mention their location here. An exception to this is Canyonlands National Park, where some extensive and well-preserved cliff ruins can be seen in Salt Canyon. Pictographs and petroglyphs—the "messages" whose meanings have yet to be deciphered—are found in hundreds of places, on the walls of the deepest and most remote canyons. Probably the best single display of these is at Newspaper Rock northwest of Monticello.

These earliest known settlers seem to have been farmers, subsisting mainly on beans, maize and squash, which they cultivated in small patches, much as the Hopi Indians do today, along canyon bottoms. Around 1300 A.D., after an extended period of drought, they all left the area and migrated south and southeastward; the Hopis, Zunis, and Rio Grande Pueblo tribes of today are probably direct descendants of the Anasazi.

The first whites to penetrate the area were the Spanish. In 1775-76 Padre Escalante and others pioneered what was later called the Spanish Trail in an important journey which took them from Santa Fe, New Mexico, to the Colorado River at what is now Grand Junction, west from there through the Uinta Basin to Utah Lake in the Great Basin, then south to the Rio Virgin. Their objective was Monterey on the California coast but with winter upon them they decided instead to return to Santa Fe. Rather than retrace their way they attempted a shortcut eastward through what is now called the Arizona Strip. They reached the Colorado River again, this time at the site of Lee's Ferry. Unable to ford the river, they found a path up over the Echo Cliffs, proceeded on along the north rim of Glen Canyon until they found a feasible crossing at a point some twenty-four miles up-river from the present Glen Canyon Dam. Once across the river the most difficult part of their travels ended; they returned through Navajo and Hopi country to Santa Fe. Escalante kept a diary and one of his companions, Miera, drew maps which later became important (if not entirely accurate) guides to the region for subsequent explorers, slavers, trappers and settlers. They had not intersected much of the canyon country but they had encircled it. Many of the names on the land are derived from these early explorations by the Spanish and their Mexican followers— e.g., Sierra La Sal, Los Abajos, the Rivers Colorado, Dolores, San Juan and Animas.

After the Spanish came the fur trappers, French and American. The first to penetrate the inner canyons and leave any record was one Denis Julien, whose name and the date "1836" have been found inscribed on the rock walls of Labyrinth, Stillwater and Cataract canyons, deep in the canyonlands.

The first official U.S. government-sponsored investigation of the canyon country was that led by Captain John N. Macomb in 1859. Starting from southwestern Colorado, he traveled westward to within a short distance of the confluence of the Green and Grand (now Colorado) rivers. In his journal Macomb wrote, "I cannot conceive of a more worthless and impracticable region than the one we now find ourselves in...!" Unable to find an easy route through this land of sculptured rock and rimmed-up canyons, he and his party returned to their base by way of the San Juan River and the present four-corners area.

In 1869 came Major John Wesley Powell and the first complete and scientific exploration of the canyons of the Green and the Colorado. He and his men were almost certainly the first to navigate the entire course of the rivers and the book *Canyons of the Colorado,* Powell's record of the voyages and travels of 1869, 1870 and 1871, is now a classic of its kind, perhaps the best as well as the best-known account of the canyon country. The names Desolation, Labyrinth, Stillwater, Cataract, Dirty Devil and Glen, as well as many others, were first given to these great central canyons by Powell. He dispelled much of the mystery surrounding the region and at the same time succeeded in describing it in a way which was not only precise but also appreciative of its sometimes grotesque, sometimes surreal beauty.

Soon after Powell's expeditions Mormon colonizers pushed eastward from their original settlements in central and southwestern Utah (along the Virgin River), following the command of Brigham Young to extend the new realm of Zion over as much territory as possible. One of the first of the new settlements was Lee's Ferry, followed by Paria (later abandoned) and other more successful establishments along the eastern foot of the high plateaus—Cannonville, Henriville, Boulder, Escalante, Fremont, Loa, Hanksville, Castle Dale and Price are towns which were founded in this period. In 1877 Moab was settled, after the failure of an earlier attempt. In the winter of 1879-80 a group of colonizers made the arduous and pioneering Hole-in-the-Rock trek from Panguitch in the Sevier Valley to what became the village of Bluff on the San Juan River. (Three babies were born on the way and all survived!) This was probably the first large-scale expedition of white people to penetrate and go clear across the most rugged and forbidding terrain— south of The Maze and The Needles—in all of the canyonlands. Their historic route was used by others for a while but eventually replaced by the easier roads to Hall's Crossing and Dandy Crossing.

From Bluff other settlers moved north to establish Blanding and Monticello, thus completing the ring of small towns around the canyon country. Of all the towns and villages I have named, only one—Hanksville—can be said to lie *within* the region; all the others are situated on the fringes. Perhaps Hite could have been considered a second, but even as the tiniest sort of hamlet it never had a continuous existence, and now, drowned out by the dam, it is only a memory. You might think, if there are no permanent towns or villages, at least there must be a number of farms and ranches within the canyon country. You would be mistaken; the only permanent human habitation deep inside this area is the Dugout Ranch on Indian Creek. There are no others, none whatsoever, within these approximately ten thousand square miles of happy hunting grounds.

This does not mean that the canyon country is a virginal wilderness, in the sense of being untouched by man. Men (and women, one hastens to add these days) have been all over, through, in and out of it for at least two thousand years. Nevertheless it remains a splendid and enchanting region, despite the intrusions. (Some of my best friends are not virgins.) As we shall see, the land has been probed and partially ravaged several times and is now being assaulted again on the largest scale ever.

Although the Mormon colonizers did not attempt to make homes within the interior of the region (aside from the exceptions named) they certainly did not ignore it. Soon realizing that they could not subsist on farming alone (even the cliff dwellers had failed at that) they began running cattle and sheep out on the mesas in summer and down in the canyons in winter. Such land cannot support much in the way of livestock—200 acres per cow is a rule of thumb —but then there was a vast amount of land and not many people trying to wrest a living from it. These ranching operations varied in size from marginal, part-time family enterprises to such gigantic land and cattle companies as the Scorup-Somerville, which controlled the entire range between the Colorado and the San Juan.

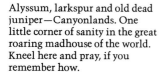

Alyssum, larkspur and old dead juniper—Canyonlands. One little corner of sanity in the great roaring madhouse of the world. Kneel here and pray, if you remember how.

After the ranchers came the prospectors and miners. Earliest of these in our territory was Cass Hite, the man who named both Dandy Crossing and the intermittent settlement which came to exist close by. Hite found a little gold dust in the banks of Glen Canyon, in the year 1883, and the word leaked out. Immediately there was a rush of miners, amateur and professional, into the area, and for the next twenty-eight years a variety of efforts was made to recover the fine placer gold from the sands and gravels of Glen Canyon. A few individuals operating shovels, sluice boxes and pans may have made a few hundred dollars from their work but more elaborate devices, such as the giant dredge built for the Hoskaninni Mining Company and operated in Glen Canyon in 1901, turned out to be complete failures: what gold existed was too fine to be separated from the silt and sand by any methods known then. By 1911 the canyon-country gold boom had petered out. Whatever gold may still be there has long since been exceeded in value, in Glen Canyon anyway, by the ever-accumulating deposits of a treasure in aluminum beer cans under the waters of Lake Powell.

The gold boom was a bust. But in the half-century that followed there was continual mineral exploration of the region, with a few modest successes here and there, as in the discovery of copper deposits near White Canyon, together with radium, vanadium and uranium in scattered locations. Uranium became a big business in 1948, starting the busiest rush to date of mineral development, one which is still continuing. Potash is being extracted from deep mines near Moab, but at great expense and dubious profit. Oil exploration, begun way back in 1898, still goes on; so far no commercial deposits have been found except at Aneth, up-river from Mexican Hat and a little outside the province of this book. Huge coal reserves, also on the fringes of the slickrock country, will soon be exploited for power production at Page and near Price.

The most recent phase in the human history and exploitation (unfortunately the two are usually the same) of the canyonlands is in the promotion of a tourist industry. Rainbow Bridge and Natural Bridges have been famous, if seldom visited, attractions to sight-seers for over half a century; the sandstone arches near Moab were made a national monument in 1933.

It was not until World War II, with all the excitement and prosperity that followed, that southeast Utah became one of the West's major tourist draws. Glen Canyon National Recreation Area and Canyonlands National Park were established by acts of Congress in the early 1960s, and once local chambers of commerce began to grasp the profit-making potential in such magic names, the appetite for road building—key to tourism—has become insatiable.

This brings my outline of the natural and human history of our region up to the present. As for the future, I am confident that its natural history will continue for a long time to come, give or take a few million years. The human story looks doubtful; but here, too, I believe that in the long run greed and stupidity will be overcome by intelligence, courage, and love.

Give or take a few million years.

# 3. Days and nights in Old Pariah

The local cowboys (they number about a dozen souls, all living in Kanab or, during working hours, in the Buckskin Tavern just across the Arizona line) call it the Pyorrhea. The Utes called it *Pah*—meaning water—and *Reah*—meaning dirty. Pah-reah, dirty water. The early Spanish explorers *españolized* the name to Paria. I like Pariah.

I'm talking about a canyon, an abandoned Mormon village, and a small perennial stream which begins near Bryce Canyon National Park and empties itself, about eighty-five meandering malingering unfairly maligned miles later, into what's left of the Colorado River at Lee's Ferry.

I haven't seen it all. I wonder if anyone has. But I've seen a great deal of it, together with Buckskin Gulch, a major tributary, and it is one of my favorite secret places in the canyon country.

Old Pariah is on the maps. You turn north off the highway onto a nice ill-graded maybe never-graded and always treacherous (especially in summer) dirt road which takes you after a few miles down among the lung-colored liver-hued bentonite hills of the Pariah Valley. The color changes continually with the variations in the weather, the declensions of sun and moon, the character of the sky.

The first thing you will see is a fake ghost town, made of second-hand lumber, built by Hollywood a few years ago for some mediocre movie. Nobody lives here. Nobody ever did. A single dirt street leads between two rows of false fronts—Saloon, Hotel, Livery Stable, Jail—all phoney as Disneyland. Some of the houses appear to rest on rock footing but check them close and you find only chicken wire and plaster. The buildings have no walls in the rear; they were built that way to allow easy access for cameras and booms. From certain angles Hollywood Pariah looks like a half-dismantled mining camp.

The boards creak and crack under the pitiless sun. When the wind blows, which is most of the time, you hear its melancholy moan through the empty wooden shells, the rattle of loose roofing, the rustle of tumbleweeds and tin cans, the swirl of red dust down the abandoned street. At midnight the town becomes still as starlight. Under the moon it appears as an intricate diagram of pale surfaces, rusted roofs, inky angles, black and impenetrable shadows full of menace, where assassins hide, where the kind of monsters never seen by light of day stand and wait—for you.

If other tourists ever find this place, it seems likely that most will mistake the Hollywood construct for the real Pariah. But the real Pariah is different.

You follow the dirt track for another mile beyond the fake ghost town till you come to the silt and sand bottoms of the Pariah River. Unless the ground is frozen it is best to park your vehicle and walk; even a four-wheel drive truck can get bogged down in the muck beyond. You wade across the river, which, except when in flood, is no more than ankle deep, and climb the left or eastern bank. Here, scattered over a mile of rocky benchland, some of it shaded by cottonwoods, are the ruins of the original town.

There are not many. Founded in 1874, Pariah never had a population greater than a hundred humans. The Mormon settlers attempted to farm the bottomlands along the stream but the frequent floods, the alkaline soil and the mineralized water made agriculture a hopeless project. In 1890 the town was deserted and no one has tried to make a permanent habitation here since. Only the cattle, property of some Kanab rancher, make any use of the shade and shelter provided by the old cabins, expertly constructed of sandstone slabs hauled from the nearby cliffs. Some of the stone lintels over the doors and windows are five feet long, four inches thick. On most are the ripple marks left by the waves of an ancient sea.

So here at Pariah we have two ghost towns, one real, one false, one of stone, one of wood, one recent, the other almost a century old, both in ruin, both sad and strange and strangely fascinating, both forgotten in a setting of surrealistic forms, ruined rock, malevolent and glowing color—a radiant wasteland.

Buckskin Gulch is the Pariah's major tributary. It is almost as long as the Pariah itself, commencing beneath the pink cliffs of the Paunsaugunt and leading nearly to the Arizona line, where it enters Pariah Canyon at grade level, 800 feet below rimrock. The last six or seven miles of Buckskin Gulch are said to be fantastically deep and narrow—"so deep and narrow," one Kanab old-timer told me, "you can see the stars in the daytime"—and it was of course this portion we hoped to see.

We began our hike—John De Puy, Judy Colella and I—off the dirt road in Houserock Valley, following the wash which we guessed was Buckskin toward the east, into a Byzantine region of sandstone domes, turrets, pinnacles, minarets, alcoves, grottoes and amphitheaters.

August: the wash was damp and firm, having recently been flooded. Pools of stagnant water stood here and there in the shade of boulders and mud banks, perceptibly evaporating—we could feel the humidity. A few clouds overhead. We passed through something like a portal, a gateway in the rock, entering a hidden basin

46

of sand, rice grass, juniper and tumbleweed enclosed within a giant bowl of almost totally nude, monolithic sandstone slickrock hundreds of feet high. A weird place, silent as the moon.

Just before the entrance to the narrows we found a heifer embedded to her belly in wet quicksand. Judging from the amount of dung beneath her tail the poor beast may have been sunk there for twenty-four hours. She'd given up struggling, anyway, though still alive. We tried to get her out but failed, the gelatinous sand flowing back as fast as we could remove it with our hands. Rather than struggle there all day we decided to go on, thinking that the heifer would be able to extricate herself by her own efforts as the quicksand dried out. We'd return the same way. Maybe.

We entered a second portal in the rock and found ourselves in a deep and narrow gulch. The walls on either side were sheer sandstone, unscalable. As we tramped deeper into this natural corridor we looked ahead, at each turning of the way, for an exit, an escape route up and out. None in sight. It must have occurred to John and Judy as well as me that it was a foolhardy thing to go into such a place in the middle of the rainy season. A cloudburst could have filled the upper reaches of the Buckskin miles away without our having any notion of it; a nice fat flash flood might already be rolling down upon us from the rear. But none of us mentioned the possibility. Perhaps from fear of reinforcing fear. We certainly had no intention of doing something rational and sensible like maybe turning around and getting the hell out of there. We slogged on, into the shadows, into the muck, into the abyss. Fortunately for fools there was no storm that day.

The pools of water got deeper, the quicksand sloppier the farther we went. We could wade the former but it became increasingly harder to slide and skate over the surface of the latter. Whenever we paused to inspect the layout ahead we discovered our feet sinking in; several times whole banks and shelves of the stuff would start to move beneath us and slough off with a splash into the water.

We were about to give up when John spotted sunlight ahead. An opening. We plowed through the last hundred yards of glue—under overhanging walls that must be four or five hundred feet high, at least—and out into a wide place in the canyon with sunshine, elbowroom, a pocket of life.

We sat down to dry out, poured the silt out of our boots, ate some lunch. Petroglyphs on the clean-cut wall; we inspected them. And a winding slit off to the right; a possible exit? We'd investigate that if we had to come back. Lacking a map, we had only the dimmest idea how far we were from the main canyon of the Pariah. Five miles? Ten miles? Onward.

The walls close in again. We tramp down a rocky hallway about five feet wide and 500 feet deep. Above us the walls are so near one another in places, and so angled, that you cannot see the sky. Nor any stars. Wedged crosswise between the walls, sometimes twenty feet above our heads, are logs—driftwood left behind by a flood. Fairy bridges. High-water marks. Some look as if the weight of a bird might dislodge them.

The gravel, pebbles and rocks disappear beneath new layers of muck. We sink in to our thighs. It *would* be a horrible way to die. Around another bend and we see long pools of muddy water waiting for us, and more shores of quicksand, and the subterranean passage winding on and on into twilight, deeper and deeper through the plateau.

At that point we all agreed to chicken out and return. Heaving ourselves through the mud back to firm terrain, we explored the side slit and found that it would go up and out.

On the surface again, baking happily in the August desert heat, we looked around for Buckskin Gulch. It was hard to make out. We stood on the rolling and canted surface of the Pariah Plateau, in the midst of a maze of naked sandstone hills. Dark cracks visible here and there; any one of them might be a canyon hundreds of feet deep. Which was Buckskin? We could only guess.

We tramped around for a while over the hills and through the dunes, found flowers and yucca and juniper and an old Anasazi chipping ground, startled a bunch of half-wild cattle, took a dip in a sandy pothole and started back. We did not descend into the canyon again; instead we followed its winding rim.

When John and Judy stopped in a shady place with seep to draw some pictures in their sketchbooks, I refilled my canteen and climbed a big sloping scarp on the north side of Buckskin. On top, among knolls and reefs and capitol domes of totally flora-less stone (I mean not a single weed, not even a blade of grass) I found the most spectacular group of potholes I have ever seen anywhere in the Southwest. There must have been thirty, all within an area of maybe ten acres, all filled with water from the recent rains. I went swimming in one which was about 150 feet in diameter and seven feet deep at the center.

Wriggling about in the water with me were swarms of tadpoles, mosquito larvae, a variety of water beetle known locally as "boatman," and the fantastic fairy shrimp, *Apus acquelius,* which looks like a miniature horseshoe crab—or trilobite. A type of fresh-water crustacean, it probably is a direct descendant of the Paleozoic trilobite which it so closely resembles. An eerie feeling, to contemplate at six-inch range this helmet-headed grope-thing with its eyeless brow, serried ranks of villus-like feet and long, forked, jointed tail—like facing the primeval here and now. I thought of the ancient slime from which it is said we are all the spawn; I thought of our earth as it will be one billion sun-years hence, when such things as these may be all the life that remains. The mind reels before such gulfs of time and mutability.

A sensational storm now brewing in the southwestern skies: amazing streamers of wind-whipped cloud, lined with lightning, fly before the sun *(oh black and scarlet banners of revolt! of hope! of free beer!).* I slid and scurried down the sculptured mounds of sandstone to my friends and the patient heifer.

Yes, on our return we found her still in place, solidly cemented in the dried and hardened sand. There went the evening. We spent two hours digging that brute out with hands and sticks. Freeing

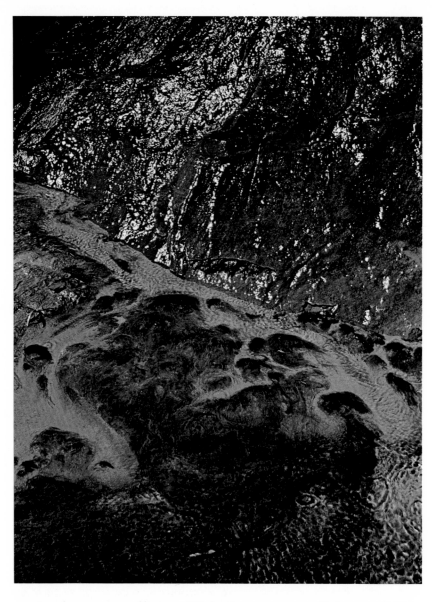

A little pool at the base of a seep.
In the desert this can be the water
of life. Of your life. Or mine.

one hoof was not sufficient. We had to excavate all four, and even then, apparently exhausted, she would not rise from the four post-holes in which our labor left her standing. Post-holes? I found a log, an old cedar fence-post as a matter of fact, and with that we levered her out of the ground. Out of the grave. She still wanted to die. So we grabbed her by the tail and hoisted her rear, then the ears and up with the head. She trembled, took a few wobbly steps like a newborn calf, halted and looked back at us. Grateful? Not that we could notice. I gave her a good boot in the hinder end to liven up her spirits. She shambled off, headed for grass and water. We went our ways.

The storm which had looked so promising blew away that evening but returned two days later in full Augustan glory. I presume that Buckskin Gulch was filled, flushed and scoured once again. I hope that cow made it.

Six weeks later I was back on the old Pariah, determined this time to go all the way. With me was Tom Lyon of Logan, wearing

50

sandals. (He carried a pair of canvas sneakers in his pack.) Starting from near the highway where the Pariah cuts through The Cockscomb, we walked for three and a half days and came out at Lee's Ferry. Tom celebrated with a swim in the icy green Colorado River. I had a beer or two.

Along the way we saw nothing but cavernous gorges, quicksand, high sheer tapestried walls of golden sandstone, marvelous patterns of light and shade on rock, water and cottonwood trees, green cottonwoods just barely beginning to turn to gold, the boggy impassable mouth of Buckskin Gulch, fresh-water seeps and springs, hanging gardens on the canyon walls, side canyons, an abandoned river channel, waterfalls and plunge pools, potshards and petroglyphs, remains of a pump station where Mormon ranchers back in the 1920s had tried to raise water from the stream to the plateau 1,000 feet above, a 200-foot natural arch, a deserted homestead, old cowboy trails, signs of deer, coyote and bighorn sheep, immense sand dunes, blazing meteors by night and a radiant sun by day and for the last ten miles an extensive panorama of plateau escarpments, sand mountains, the Echo Cliffs and the original sky.

Tom's sandals gave out about half-way. It was a noble experiment. My back ached from carrying twenty pounds too much dehydrated enchilada and freeze-dried bouillabaisse, an ignoble experiment; the trouble was we walked too fast and didn't eat enough. We should have spent ten days on that trip. I believe I'd be willing to spend ten percent of the rest of my life in the canyons of Pariah—if they leave them alone. Twenty percent. Of the canyons which still remain in Utah's canyon country (the most beautiful of all now being kept in liquid storage) I know of only a few which match or surpass Pariah.

Luckily for those who prefer their canyons natural, it seems unlikely that Pariah will ever be subjected to much official improvement or motorized invasion. Vertical walls, the narrow canyon floor, frequent and vigorous floods make any kind of road unfeasible. Because of soft sand, spongy gravel, boulder jams and the lengthy stretches of quicksand, anyone who tries to get through here on a totegote or similar frivolous device will find himself toting his gote on his back for most of the forty-two miles; even a saddle horse would be a tough proposition. Donkeys and burros, maybe. The Bureau of Land Management plans to install "a few discreet signs" within the canyon to indicate points of interest to those who stumble in here by accident (wild-eyed with terror, seeking a public telephone?), but these will be removed by the first flood to come along—or by the first thoughtful hiker.

Good old American know-how: a little more enterprise and we may yet be free....

# 4. Fun and games on the Escalante

The Escalante is a small river which flows from the plateaus of south-central Utah into what was the Colorado River near its junction with the San Juan River. By any but desert standards it would be called a creek. Or crick. During most of the year it runs shallow, not more than a foot deep, but in the spring and late summer, swollen with snow-melt or cloudbursts, it looks more like a real river and bears a heavy load of silt and sand.

Armed with these abrasives and aided by a liberal allowance of time, the little Escalante River has carved a deep, winding and dramatic canyon through the massive and monolithic sandstone formations. Toward its deeper end, near the currently submerged channel of the Colorado, the walls of the Escalante's main canyon are more than 1,500 feet high—and sheer, vertical or overhanging, slick as the wall of your living room, with only the smallest of niches for such things as cliff swallows, canyon wrens, owls, hawks and bats.

In the flat sunlight of midday the mighty cliffs appear buff-colored, a pale auburn, but at morning and evening when the sun's rays come slanting in at a low angle the rock takes on an amber tone, with a glow like the bead of good bourbon. At sundown the coloring deepens still more and smolders—you can feel the heat—through the twilight in all the hues of hot iron cooling off.

There is much more to the Escalante and its surrounding 500,000 acres of de facto wilderness than the central gorge. The Escalante is a system of canyons, dozens of canyons, all feeding into the river, each of them rich in marvels: intricate detail in water and stone, plants and animals, light and shade and color, solitude and stillness. Most of the tributary canyons contain seeps, springs, perennial streams. There are waterfalls and pools, and in some of the canyons great natural stone arches and bridges, such as Gregory Bridge (presently submerged by Lake Powell, but waiting), Stevens Arch, Broken Bow Arch, Hamblin Arch.

Above and between the canyons is the slickrock benchland, that weird world of hills, holes, humps and hollows where, they say, the wind always blows and nothing ever grows. But even here, in what looks at first sight like nothing but naked rock, there are pockets of life. In the wind-drilled potholes you will see, after a rain, the resurgence of living things: tadpoles, mosquito larvae, fairy shrimp, threadworms and water beetles. Between rains these natural waterholes go dry but the life is still there, buried under the sediment in the form of eggs, spores and seeds, and sometimes even adult, estivating toads. Dove feathers and coyote scat offer evidence that the water is known to more than insect and amphibian.

A variety of desert plants grow in the sandy basins among the knobs and pinnacles, despite the local mythology. Plants such as juniper, yucca, prickly pear, sand sage and chamiso may not be transmutable, through the stomachs of cows, into money in the bank but they are things of interest and beauty and therefore of value all the same.

Except for the sixty-mile dirt road from the town of Escalante to the dead-end point called Hole-in-the-Rock, there are no permanent roads within the Escalante area. Some older maps show a jeep trail crossing the Escalante at Harris Wash but this is seldom used. The last time I went down there, in the spring of 1970, some stockmen had strung a fence across it. Oil companies have also bulldozed a few temporary roads on the benchlands above the canyons but these are soon made impassable by erosion, although they leave scars on the land which will take a long time to heal. Except from the air, however, they are not noticeable. There are no roads of any kind down in the Escalante and its side canyons nor even any man--made trails. A few old corrals and cabins on the benches above the canyons add a picturesque note to the scene; they do not detract from its primitive character. The Escalante may not be a *completely untouched* wilderness—where on earth is there any such thing any more?—but it is the closest thing to it that still remains in southeast Utah.

Best of all, the Escalante country belongs to *us*. It lies entirely within the public domain, and is therefore the property not of land and cattle companies, not of oil and mining corporations, not of the Utah State Highway Department or any Utah Chamber of Commerce, but of *all* Americans. It's *our* country.

Or should be. It's supposed to be.

There are those who have other ideas. The local chamber of commerce, for example. Representing chiefly the motel owners and gas-station operators of Garfield County (total population: 3,200), in which most of the Escalante wilderness area lies, the county C. of C. tends to look upon our lands as *their* property. Psychologically this is understandable; many of the local people have lived here all their lives and their ancestors, beyond question, were the first to settle here. After the Indians, I mean. A few of the local people, the cattle growers, depend upon the public lands as a cheap source of forage for their herds, although the cattle industry is now of trivial importance in the economy of southeast Utah (see Author's Notes).

When the State Highway Department proposes to build a paved

highway through the heart of the Escalante area it is to be expected that local chambers of commerce will support the project. According to traditional business doctrine, road construction of any kind, no matter how destructive or foolish, is bound to bring in some sort of revenue, for a time, to local business. Always providing, of course, that the costs of the project are borne by the public at large—by the rest of us.

The Escalante road project is as follows: from Bullfrog Basin, a small marina on the Glen Canyon Dam reservoir, the road would be built southwesterly over or somehow through the Waterpocket Fold, across the canyons of the Escalante near the mouth of Coyote Gulch, and from there up over the sand and slickrock terrain to a junction with the present Hole-in-the-Rock road. (See map.)

The Escalante proposal is actually only a segment of a much more grandiose scheme. What the Utah Highway Department really wants to do is build a highway from the Colorado border northeast of Moab all the way to a shabby little "development" called Glen Canyon City close to the Arizona border at Page. The proposed route would lead from Cisco to Moab along the river, from there up over the plateau north of Dead Horse Point and Canyonlands National Park, down into Taylor Canyon and across the Green River below Bowknot Bend, thence southwestward past the fragile and so-far unspoiled wilderness of The Maze, the Standing Rocks and Black Ledge to connect with Utah State Road 95 near Hite Marina. From there the so-called "Lake Powell Parkway" would coincide with the present paved road to Bullfrog Basin. At that point would begin the Escalante segment, followed by more new road through heretofore untouched wilderness over the Grand Bench south of the Kaiparowits Plateau to Glen Canyon City.

Excepting Moab, the road would not come near any existing Utah town; for most of its distance it would pass through what is now *totally uninhabited* wildland. Sole purpose of the road, even in the eyes of its proponents, is the promotion of mass motorized tourism, in the manner of Great Smokies and Yellowstone, for southeast Utah.

Since the route as proposed follows no natural line of travel but goes directly against the grain of the land, the cost both in dollars and in scenic values would be enormous. Portions of the highway would cost an estimated $2 million per mile for construction; there is no way to measure the loss entailed in the destruction and pollution of the landscape. And who would pay these costs? The Utah State Highway Department? The Utah chambers of commerce? Not bloody likely. Although resentful of any suggestions from those whom they consider "outsiders," Utah businessmen do expect us to subsidize their economic development schemes. They hope to pay for this road with federal funds—our money. Your money.

Back to the Escalante affair. Rather than attempt to promote their entire highway project at once, the Highway Department is pushing for it in sections, hoping in that way to stir up less opposition. But in this the department has failed. The Escalante road proposal has aroused determined resistance from Utah conservationists, par-

A slickrock garden in Capitol Reef. There are many such places in our kind of country—but are there enough? Or will we have to put this under glass?

ticularly those represented by the Wasatch Mountain Club (no affiliate of the Sierra Club). But conservationists are still a minority in the state of Utah, where the road-building mania dies hard, and so their opposition to the proposed road has led to a lively, sometimes nasty, controversy.

The conservationists and wilderness advocates (sometimes called "wild conservatives" by the local press) contend that funds for economic development in southeast Utah should be spent on the improvement of already-existing roads in the area, many of which are yet unpaved. Some argue that immediate human needs, such as medical facilities, better schools, and pollution controls, should be given priority over road building. Others suggest that federal funds be used to buy up grazing permits, freeing the Escalante wilderness from the domestic cattle that now infest the area. The Wasatch Mountain Club and its allies also urge official and permanent wilderness classification for the canyons in order to protect them once and for all against threats of highway construction and mineral exploitation.

Such counterproposals have led inevitably to counter-reactions as the dialectic of controversy proceeds. As might be expected, the promoters of the new highway are in no way mollified or appeased by the suggestion that present roads be improved first. They want that anyway, in addition to new roads where none now exist. They see new roads, any new roads anywhere, if financed with federal funds, as "a shot in the arm for the local economy" and therefore intrinsically good. No matter if the result is a shot in the head for older and more basic human values. Thus the road builders and regional businessmen are bewildered and outraged when they see their hopes threatened by the inexplicable chicanery of outside agitators. Consider the following documents, the first published as an advertisement in the *Salt Lake City Tribune* by the (town of) Escalante Chamber of Commerce:

WE WANT TO SHARE, NOT HOARD...MUST WE ALSO HAVE A WILDERNESS?...The Protectionists of the Wasatch Mountain Club and the Sierra Club are proposing an even larger Wilderness Area for the Escalante region. It would prohibit the building of new roads. Only trails are allowed in Wilderness Areas...These Protectionists are totally unrealistic, unjust and ruthless. They would kill the economy of a country to gratify their passion for exclusive use of territory that has scenic values. THEY WANT SINGLE USE. We want Multiple Use of the public lands. By the way, where do they get their great political power?....

Here is another, excerpted from a circular distributed by the same chamber of commerce:

TO ALL FAIR-MINDED PERSONS FROM THE RESIDENTS OF EASTERN GARFIELD COUNTY: The Wasatch and Sierra Clubs want to keep the country for the exclusive enjoyment of those who are able to take expensive tours in it. They get satisfaction from the thought that only they are able to appreciate natural beauty...

This is natural wilderness country. It is in no danger of destruction or defacement...There will always be plenty of wilderness here for everyone.

55

In order to achieve the "Protection" they clamor for, these clubs have asked for the creation of a 600,000 acre Wilderness Area...This would destroy the basic industry of Garfield County, that of cattle grazing.... There could be no roads, no oil or mineral development....(see Author's Notes).

An official of the Escalante Chamber of Commerce also wrote to the Wasatch Mountain Club and its Escalante Wilderness Committee, condemning the club's "diabolical plot to lock up this natural resource into a Wilderness Area to which only a few people would have access" and threatening to seek an injunction against the committee's use of the name Escalante. When Jack McLellan, committee chairman, requested a meeting with Escalante Chamber of Commerce members he received the following reply:

Dear Mr. McLellan:

As I promised, I brought the matter of your proposed visit to us before a meeting of the Chamber of Commerce last evening. The vote was unanimously against asking you to come.

W. M. Christensen
Chairman, Publicity Committee
Escalante Chamber of Commerce

Typical of regional attitudes are those expressed in this editorial which I reprint here exactly as it appeared in the *Garfield County News* on May 7, 1970:

BUTTERFLY CHASERS GRAB AGAIN

The Butterfly Chasers are after us again.

The People-Haters are trying to engineer another big steal of Garfield County Land.

One of those paper organizations, a basically meaningless "write your congressman" group of conservationists with a high-salaried drumbeating executive vice-president, secretary or such, who is out to earn his keep and to show the muscle of a blivvit of form letters is getting greedy again.

They want to grab 640,000 acres of Garfield County Canyon country and set it aside as a preserve for the vacationing millionaire—no one else would be able to afford the time and outfit needed for a foot-safari into their proposed wilderness of Eastern Garfield County.

They want the entire Escalante River Canyon set aside for the finicky few.

They particularly don't want roads—no, no, no—that would let people in to see this wonderful country, and you must remember, they hate people.

The old ladies in tennis shoes, semi-senile retirees, googly-eyed bleeding hearts, bow-wows and others of this ilk will now stir up a blivvit of form letters to snow their congressmen—and as every congressman knows rocks have more rights than people—maybe not as many votes—but more rights.

Most of the people who will now begin bombarding their congressional representatives haven't climbed a mountain in years. Not one out of a thousand who will bleed at every pore in defense of this wide wonderful wilderness will ever see it.

A question of this sort should be considered by the people at the scene. No steal of land within the boundaries of any subdivision of the nation should be considered without first consulting the people who are there.

With too much of the area under the mishandling of Uncle Stupids bumbling bureaucrats absentee-landlordism we can look forward to years of such maneuvers.

The bitterness of these attacks on the wilderness proposal has been accompanied by flurries of letters—"blivvits"—to state and local newspapers assailing conservationists as "land grabbers," "terrible militant conservation snobs," "wild conservatives," and "wilderness fanatics preaching sterile conservationism." When one defender of the wilderness rose in a public hearing to state his reasons for opposing the trans-Escalante highway project he was denounced as "not only selfish but unethical—even wicked!" Wasatch Mountain Club proposals are routinely referred to as "wilderness threats."

Even innocent strangers have been caught up in the hard feelings that the controversy has engendered. Here is a letter that appeared in *The New York Times:*

To the Editor:

Readers whose vacation itinerary this summer may include southern Utah might appreciate a timely warning gleaned from the unfortunate experience I had there recently. While in the area of Escalante, my car was vandalized and robbed.

A businesswoman in town assured me, with smug satisfaction, that I had been so victimized because my car carried the sticker of a national conservation group. Anyone traveling in southern Utah is well advised

Anderson Bottom, along the Green. A great good place. A man could whittle away his life down here and never lose a minute. Love it or leave it alone.

to remove or hide anything identifying him as a conservationist, for he is persona non grata with those vocal locals who have whipped up feeling against a conservationist proposal to protect the magnificent Escalante Canyon country....

Also suspect and considered fair targets are would-be visitors carrying camping equipment and particularly hiking gear. Campers are frowned upon because they supposedly will not drop enough money in the area, and hikers and backpackers are beyond the pale.

Nancy E. Williamson<br>Trenton, N.J.

Opposition to the projected road and support for wilderness classification for the Escalante area continue in the state of Utah, despite the local press, commercial interests, highway builders and state politicians. In fact, a former president of the Escalante Chamber of Commerce, Kenneth Sleight, has announced his opposition to the road in the form of a "Special Report" of his own, in which he writes:

The most destructive road proposed is the one which would lead from Bullfrog Basin to Hole-in-the-Rock [and thence] southward to the lower Lake Powell area. The reasoning given for this road is that many people, especially boaters, would prefer to return by another route than the one by which they came.

This road would lead through the heart of the few wilderness regions which yet remain. This includes the geologically unique Waterpocket Fold, the deep and enchanting canyons of the Escalante, and the towering and rugged benchlands which circle the Kaiparowits Plateau. This region is noted for its variety of topographic forms. In 1936 much of this area was proposed as a national park by Harold L. Ickes, then Secretary of the Interior. The character and the beauty remain unchanged.

The Governor of Utah has applied to the federal government for funds with which to build this road. This was done under the provisions of The Economic Opportunities Act (Poverty Bill) in "behalf" of Garfield County. This was done in spite of the fact that (1) Garfield County did not request such action, (2) the population of the county lies far removed from the proposed road location, (3) nearly all of the proposed road location is outside of the county, and (4) Garfield County citizens have made repeated requests for aid on their own local, unimproved roads....

Members of the Wasatch Mountain Club and representatives of national conservation groups attended the public hearings that were held in three small southern Utah towns—the atmosphere of these assemblies can easily be imagined—and the final hearing in Salt Lake City, where the conservationists present outnumbered the developers in the ratio of four to one. It remains to be seen whether such hearings are more than a formality—a "shouting contest" at which public officials (and the powers behind them) listen politely to citizen protest and then go ahead and do whatever they'd planned to do in the first place (see Author's Notes).

At the height of the controversy, I went down into the canyons of the Escalante. I camped, alone, for three days at a place near the junction of Coyote Gulch and the Escalante River, the very spot which will be blasted, bulldozed and bridged if the highway project is allowed. I saw that great window in the rock called Stevens Arch. In the sand near the stream I found the track of a mountain lion. I drank at a spring in the canyon wall under a hanging garden of maidenhair fern, scarlet monkey flower and pale yellow columbine. Through the sweet stillness I heard, now and then, the song of a canyon wren, the scream of a hawk.

No traffic. No stink of auto fumes or screech of brakes, no clatter of beer cans and pop bottles, no crowds, no ranger-police, no KEEP OFF, KEEP ON or KEEP OUT signs, no transistor radios, no old socks or wine bottles in the clear water, no smell of burning garbage from a trash-jammed fireplace.

And how did I get to such a place of marvels, such an Eden and Paradise deep in the wild? Why I got there in the same way as the Indians did, as the pioneers did, as a troop of little boy scouts had the year before, as a pair of blue-haired ladies from Wisconsin were preparing to do when I came out: in the best way, the easy way, the only way.

I walked.

The trail begins in Hurricane Wash. Hurricane Wash *is* the trail. The way is well trodden by the hooves of cattle, the boots of hikers. If you're out of luck you might even find the track of some clown on a motorbike going down into the canyons before you. But he won't get far: as in Buckskin Gulch and the Paria, the main canyon of the Escalante is much too rocky, rugged and quick for such contraptions. I passed a slab of sandstone on which a troop of boy scouts had petroglyphed their presence only a year before. That takes care of the argument that hiking in the Escalante is a pastime for the idle rich.

Gradually the wash deepens, becomes a little canyon. There is some plant life, not much: the canyon floor is nothing but dry sand and rock. The walls close in, forming a channel of sand and sandstone, then widen again as the canyon sinks deeper. The first signs of water appear: wet sand, alluvial mud banks, pickleweed, wire grass and clumps of willow. Cow dung and flies, of course. Deer prints and cliff rose. Dried algae on the rocks, then tiny stagnant pools with water-skippers, dragonflies and swarms of wriggling larvae.

In a deep-shaded undercurve of the wall I came to the first seep and a perceptibly flowing rill of water. One could drink here, if necessary, although the stink of cattle discourages the thought. I trudged on without pausing; although hot and thirsty I had a gallon of water in my pack, untouched. Besides, I wanted to see how far I could go in tolerable comfort without a drink. In that way I'd find out how much water I'd need for the hike out, back to the car.

Hurricane Wash cuts deeper and deeper into the rock and the walls become much higher than the canyon is wide. The little pools are joined to one another now by a trickling rivulet which oozes out of the mud and sand and slides, barely moving, over the slick rock. I passed some cottonwoods growing tall and slender in this deep shadowy canyon. More thickets of willow and tamarisk, many tracks of deer, cattle, smaller beasts.

From somewhere ahead, around the next bend, I could hear the alluring sound of rushing water. I hurried on and found that the wash now joined a much bigger canyon. Here were sand beaches, groves of giant cottonwoods and a clear-flowing creek—not a stream but a creek—which issues, brisk as a bee, from a gap in the walls.

I sat down in the shade and studied my map. I'd come about five miles and had now reached Coyote Gulch, one of the major side canyons of the Escalante. The Escalante itself was still seven or eight miles away, as nearly as I could judge from the elaborately meandering course of Coyote Gulch. I rewarded myself with a good drink of water, switched from boots to tennis shoes and went on down the gulch, slogging through the water when necessary. The stream-bed itself makes the best trail, most of the way.

Rock and water, huge cliffs and delicate details. On my right I first heard, then saw, a thread of water falling from a seep in the undercut wall of the canyon. I stepped over to it, held out my cup and listened to water tinkling into tin—most musical of desert sounds. Over my head, moist and dripping with a fine spray, was another rock-wall garden of ferns and orchids, ivy and columbine. Here would be a good place to tank up, fill a canteen or two, before beginning the walk out.

I went on.

The canyon curves deeply to the left and right, sinuous as a snake, no more willing to follow a straight line than is anything else true and beautiful and good in this world. The walls curve not only laterally, with the winding of the stream-bed, but also vertically, parallelling each other like the surfaces of a ball and socket. Where a wall is deeply undercut it forms something like the inside of a half-dome. Standing inside one of these alcoves, which may be hundreds of feet high, you will not be able to see the sky at all; the light is reflected and refracted from the opposite canyon wall, creating a strange golden ambience within the chamber.

Because of the looping course of the canyon, with high walls which often shut out the sun, you can never determine with any precision what direction you're following. In fact your points of orientation are reduced to a simple pair: upstream and downstream. Which is really all you need anyway.

I came to Jacob Hamblin Arch. Weathered through the neck of a sandstone fin that juts out into a bow of the canyon, this arch will someday become a natural bridge, when the creek below completes the work of chiseling a shortcut through its own meander.

The way up through the arch looked rough and rocky; with the heavy pack on my back I preferred not to try it. Instead I followed the creek on its way around the gooseneck, passing beneath one of the largest half-domes I had ever seen. Imagine the Hollywood Bowl expanded to ten times its present size. The apex of this structure must be at least 500 feet above the stream-bed.

The last of the afternoon sunlight had long since vanished from the upper walls. I walked through lavender twilight, through the sounds of flowing water, rasping toads, swaying willows, the papery rustle of cottonwood leaves. It was time to make camp if I wanted to cook before dark, but the charm and magic of this canyon were so great I didn't want to stop. Each turn in the walls promised some

The writing on the wall. A symbiotic union of fungi and algae, lichens like alchemists transmute stone into soil, given a little time.

new delight; exploring such a place is like exploring the personality of a new friend, a new love.

But, since I would not stop for dark, the dark stopped for me, surrounded me. On a shore of sand well away from the cattle paths I made my bed and prepared supper in the starlight. Cottonwood makes poor fuel, burns too fast and smoky; I scrambled up the talus under the cliff where I had earlier noticed scrub oak and brought back an armload of *real* wood. Over the red coals of oak I hydrated my dehydrated meat and vegetables, stirred up a quick goulash. Well seasoned with salt and pepper and blow-sand, nothing could have tasted better.

In the morning I went on down the gulch, leaving my pack behind, carrying only enough raisins and jerky for lunch. I was now below the seep-line in the overlying Navajo sandstone and the stream grew larger, fed by the many springs. I climbed around a gap in the rock that had once been a natural bridge, passed several small waterfalls, and came to a natural bridge that was still standing.

This is a young bridge, geologically speaking, about big enough to drive a school bus through, with plenty of room for enlargement. There were the mud homes of cliff swallows on the inside, and under the base of the outer buttress I saw the gray smudge of an old campfire. Some hobo might have camped here once (having wandered far from the steel trail) — I probably knew the man. Or maybe it was Everett Ruess.

I walked under the bridge, feeling the sensuous pleasure of moving through a wall of stone, wading the stream that made the opening, standing in shadow and looking back at the upstream canyon bathed in morning light, the sparkling water, the varnished slickrock walls, the fresh cool green of the cottonwoods, the pink and violet plumes of tamarisk.

From the cliffs far above I could hear the clear falling notes of a canyon wren — characteristic song in this land of stone and stillness.

What would it be like to *live* in this place? Could a man ever grow weary of such a home? Someday, I thought, I shall make the experiment, become an ancient baldheaded troglodyte with a dirty white beard tucked in my belt, be a shaman, a wizard, a witch doctor crazy with solitude, starving on locusts and lizards, feasting from time to time upon lost straggler boy scout.

Madness: of course a man would go mad from the beauty and the loneliness, both equally mysterious. But perhaps it would be — who can say? — a kind of *blessed* insanity, like the bliss of a snake in the winter sun, a buzzard on the summer air.

The canyon grows bigger, wider, wilder as it descends by jump-off and cataract toward the Escalante. At one place a vast section of the north wall has collapsed quite recently, perhaps within the past century, and tumbled in blocks the size of boxcars to the canyon bottom. The rock-fall is clean, sharp-edged, free of all plant life, even lichens, and the floods have not yet had time enough to round off and polish the broken slabs that obstruct the creek. Here the footing is tricky, impossible for cattle and horses; I found steps chipped in the sandstone above the fall where somebody — Indian? Mormon cowboy? Ken Sleight? — had made a by-pass for his horse.

More cascades, some of them fifteen to twenty feet high. At one point the only route is down a log leaning against a shelf of stone on the canyon wall. Here the walls must be close to 1,000 feet high on either side and the sky no more than a narrow strip of blue.

Around one more bend, hearing the soft steady roar of floodwater, and I came to the master stream, the Escalante River.

The water was reddish-brown that day, about fifty feet wide, knee-deep in the main channel as I discovered when I waded in. The current strong and swift. Since the river at this point filled its bed from wall to wall I had no choice but to wade through if I wanted to go on. I headed *up* the river; at least that way I'd have a good chance to make it back to the mouth of Coyote Gulch if the river should suddenly rise.

The going was hard against the current. I could feel the sand giving way beneath my feet at every step. I went only far enough up the river to see Stevens Arch and to reach the mouth of Stevens Canyon beyond. I had hoped to explore Stevens but the day was already more than half over; we were running short of light. I walked a mile or so up Stevens Canyon—a rare, secret, lovely place it seemed—and then, most reluctantly, turned back, waded the river, and trudged the long and winding rock-and-mud sand-and-water trail up the meanders of Coyote Gulch. A second time I cooked my supper in the dark but it didn't matter. I was tired, hungry, happy.

The next day, before returning up Hurricane Wash to my car, I decided to have a look at the world that lay above and between the canyons. Upstream from the confluence of the wash and Coyote Gulch I found an egress from the canyon by scaling a rounded hummock of sandstone.

I kept climbing until I topped out on a bare ridge about halfway between Coyote Gulch and Escalante Canyon. From there I could see a great deal of the world: not only the dark gashes of the canyons below but also a stretch of the Waterpocket Fold, the snowy dome of Navajo Mountain, the Straight Cliffs of Kaiparowits, and other mesas and plateaus east of the Colorado River, some of them I suppose fifty miles away by line of sight.

Nearby was the equally interesting terrain of the slickrock boondocks—naked sandstone shaped by ages of weathering and erosion into the science-fiction landscape of fins and pinnacles, knobs, nodes and knolls, potholes and hollows. Some of the potholes contained old rain water which I sampled, along with a sampling of the bugs, beetles, dead flies, worms and smaller things, some dead, some alive, which flourished in the water. Spring flowers were blooming in the sandy basins: cactus, cliff rose, paintbrush, verbena, princess plume, purple penstemon, globemallow, scarlet penstemon, purple beeplant (also known as Cowboy's Delight) and others.

I'd neglected to mark my trail and so spent a couple of hours searching for the way back down into Coyote Gulch. All of the sandstone hills look dismayingly alike, at least when you're lost, and my first two approaches to the canyon ended at rimrock. I was rimmed up. Nor could I simply follow the rim until I hit my spoor, for it—the rim—is broken at numerous points by precipitous "hanging canyons" too wide to jump, too deep and sheer to descend into. I had to backtrack and circle, once, twice, a third time, until I chanced upon the one and only route down. If I were the John Muir I'd like to be I would have spent the night up there.

Down in the canyon I returned to my pack, filled my belly with seep water, had something to eat. Little hog-nosed bats with translucent wings flickered over my head in the evening light, making clicking sonar noises. A swarm of gnats performed their ritual molecular dance in the air before my face. A bat swept through the swarm, scooping dozens away forever. The gnats closed ranks and carried on the dance, indifferent to disaster. Should one despise their passive fatalism—or envy their nonchalance?

In dusk and desert music I walked up peaceful Hurricane Wash toward my wheels and that road which leads—to where? not back home: who can speak of home anymore? who can say he has not forsaken home?—but back to where we all are now.

# 5. The damnation of a canyon

There was a time when, in my search for essences, I concluded that the canyonland desert has no heart. I was wrong. The canyonlands did have a heart, a living heart, and that heart was Glen Canyon and the golden, flowing, wild Colorado River.

In the summer of 1959 a friend and I made a float trip in little neoprene boats down through the length of Glen Canyon, starting at Hite and getting off the river near Gunsight Butte—The Crossing of the Fathers. In this voyage of some 150 miles and ten days our only motive power—and all that we needed—was the current of the Colorado River.

In the summer and fall of 1967 I worked as a seasonal park ranger at the new Glen Canyon National Recreation Area. During my five-month tour of duty I worked at the main marina and headquarters area called Wahweap, at Bullfrog Basin toward the upper end of the reservoir, and finally at Lee's Ferry down-river from Glen Canyon Dam. In a number of powerboat tours I was privileged to see almost all of our nation's newest, biggest and most impressive "recreational facility."

Having thus seen Glen Canyon both before and after what we may fairly call its damnation, I feel that I am in a position to evaluate the transformation of the region caused by construction of the dam. I have had the unique opportunity to observe firsthand some of the differences between the environment of a free river and a power-plant reservoir.

One should admit at the outset to a certain bias. Indeed I am a butterfly chaser, a googly-eyed bleeding heart and a wild conservative. I take a dim view of dams; I find it hard to learn to love cement; I am poorly impressed by concrete aggregates and statistics in the cubic tons. But in this weakness I am not alone, for I belong to that ever-growing number of Americans, probably a good majority now, who have become aware (however inarticulately) that a fully industrialized, thoroughly urbanized, elegantly computerized social system is not suitable for human habitation. Great for machines, yes. But unfit for people.

Lake Powell, formed by Glen Canyon Dam, is not a lake. It is a reservoir, with a constantly fluctuating water level—more like a bathtub which is never drained than a true lake. As at Hoover (or Boulder) Dam, the sole purpose of this impounded water is to drive the turbines that generate electricity in the powerhouse at the base of the dam. Recreational benefits, while substantial, were of secondary importance in the minds of those who conceived and built this dam. As a result the volume of water in the reservoir is continually being increased or decreased according to the requirements of the Basin States Compact and the power-grid system of which Glen Canyon Dam is a component.

The rising and falling water level entails various consequences. One of the most obvious, well known to all who have seen Lake Mead, is the "bathtub ring" left on the canyon walls after each draw-down of water, or what rangers at Glen Canyon call the Bathtub Formation. This phenomenon is perhaps of no more than esthetic importance; yet it is sufficient to dispel any illusion one might have, in contemplating the scene, that you are looking upon a natural lake.

Of much more significance is the fact that plant life, because of the unstable water line, cannot establish itself on the shores of the reservoir. When the water is low plant life dies of thirst; when high it is drowned. Much of the shoreline of the reservoir consists of near-perpendicular sandstone bluffs where very little flora ever did or ever could subsist, but the remainder includes bays, coves, sloping hills and the many side canyons where the original plant life has been drowned and new plant life cannot get a foothold. And of course where there is little or no plant life there is little or no animal life.

The utter barrenness of the reservoir shoreline recalls by most exemplary contrast the aspect of things before the dam, when Glen Canyon formed the course of the untamed Colorado. Then we had a wild and flowing river lined by boulder-strewn shores, exquisite sandy beaches, thickets of tamarisk and willow, and glades of cottonwoods.

The thickets teemed with songbirds: vireos, warblers, mockingbirds and thrushes. On the open beaches were killdeer, sandpipers, herons, ibises, egrets. Living in grottoes in the canyon walls were swallows, swifts, hawks, wrens and owls. Beaver were common if not abundant: not an evening would pass, in drifting down the river, that we did not see them or at least hear the whack of their flat tails on the water. Above the river shores were the great recessed alcoves where water seeped from the sandstone, nourishing the semitropical hanging gardens of orchid, ivy and columbine, with their associated swarms of insect and bird life.

Up most of the side canyons, before damnation, there were springs, sometimes flowing streams, waterfalls and plunge pools—the kind of marvels you can now find only in such small-scale remnants of Glen Canyon as the Escalante area. In the rich flora of these laterals the larger mammals—mule deer, coyote, bobcat, ring-tailed cat, gray fox, kit fox, skunk, badger and others—found a home. When the river was dammed almost all of these things were lost. Crowded out—or drowned and buried under mud.

The difference between the present reservoir with its silent sterile shores and debris-choked side canyons, and the original Glen Canyon, is the difference between death and life. Glen Canyon was alive. Lake Powell is a graveyard.

For those who may think I exaggerate the contrast between the former river canyon and the present man-made impoundment, I suggest a trip on Lake Powell followed immediately by another boat trip on the river below the dam. Take a boat from Lee's Ferry up the river to within sight of the dam; then shut off the motor and allow yourself the rare delight of a quiet, effortless drifting down the stream. In that fourteen-mile stretch of living green, singing birds, flowing water and untarnished canyon walls—sights and sounds a million years older and infinitely lovelier than the roar of motorboats—you will rediscover a small and imperfect sampling of the kind of experience which was taken away from us all when the oligarchs and engineers condemned our river for purposes of their own.

The effects of Glen Canyon Dam also extend downstream, causing changes in the character and ecology of Marble Gorge and Grand Canyon. Because the annual spring floods are now a thing of the past, the beaches are becoming overgrown with brush, the rapids are getting worse where the river no longer has enough force to carry away the boulders washed down from the lateral canyons, and the litter and dung left behind by ever-increasing human travel, formerly flushed out each spring, now accumulate at a rising rate from season to season.

Lake Powell, though not a lake, may well be as its defenders assert the most beautiful reservoir in the world. Certainly it has a photogenic backdrop of buttes and mesas projecting above the expansive surface of stagnant waters where cabin cruisers play. But it is no longer a wilderness. It is no longer a place of natural life. It is no longer Glen Canyon.

The defenders of the dam argue that the recreational benefits available on the surface of the reservoir outweigh the loss of Indian ruins, historical sites, wildlife and wilderness adventure. Relying on the familiar quantitative logic of business and bureaucracy, they assert that whereas only a few thousand citizens ever ventured down the river through Glen Canyon, now millions can—or will—enjoy the motorized boating and hatchery fishing available on the reservoir. They will also argue that the rising waters behind the dam have made such places as Rainbow Bridge accessible by powerboat. Formerly you could get there only by walking. (Six miles.)

This argument appeals to the wheelchair ethos of the average American slob. If Rainbow Bridge is worth seeing at all, then by God it should be easily, readily, immediately available to everybody. Why should a trip to such a place be the privilege only of those who can walk six miles? Or if Pike's Peak is worth getting to, then why not build a highway to the top of it so that anyone can get there? Anytime? Without effort? Or as my old man would say, "By Christ, one man's just as good as another—if not a damn sight better."

The scrubbiest little vegetable may produce the biggest and boldest flower. God lives and his name is *Carry On.*

Or as ex-Commissioner Floyd Dominy of the U.S. Bureau of Reclamation pointed out poetically in his handsomely engraved and illustrated brochure *Lake Powell Jewel of the Colorado* (produced by the U.S. Government Printing Office at our expense): *There's something about a lake which brings us a little closer to God.* In this case, Lake Powell, about 500 feet closer. Eh, Floyd?

It is quite true that the flooding of Glen Canyon has opened up to the motorboat explorer parts of side canyons that formerly could be reached only by people able to walk. But the sum total of terrain visible to the eye and touchable by hand and foot has been greatly diminished, not increased. Because of the dam the river is gone, the inner canyon is gone, the best parts of the numerous side canyons are gone—all hidden beneath hundreds of feet of polluted water, accumulating silt, and mounting tons of trash. This portion of Glen Canyon—and who can estimate how many cubic miles were lost? —*is no longer accessible to anybody.* And this, do not forget, was the most valuable part of Glen Canyon, richest in scenery, archeology, history, flora and fauna.

Not only has the heart of Glen Canyon been inundated but many of the side canyons above the fluctuating waterline are now rendered more difficult, not easier, to get into. This because the debris brought down into them by desert storms, no longer carried away by the river, must unavoidably build up in the area where flood meets reservoir. Narrow Canyon, for example, at the head of the impounded waters, is already beginning to silt up and to amass huge quantities of driftwood, some of it floating on the surface, some of it half afloat beneath the surface. Anyone who has tried to pilot a motorboat through a raft of half-sunken logs and bloated cows will have his own thoughts on the accessibility of these waters.

Hite Marina, at the mouth of Narrow Canyon, will probably have to be abandoned within ten or twenty years. After that it will be the turn of Bullfrog Marina. And then Rainbow Bridge marina. And eventually, inevitably, whether it takes ten centuries, or only one, or maybe a deal less, Wahweap. Lake Powell, like Lake Mead, is foredoomed sooner or later to become a solid mass of mud, and their dams—both waterfalls. Assuming, of course, that either one stands that long.

Second, the question of costs. It is often stated that the dam and its reservoir have opened up to the many what was formerly restricted to the few, implying in this case that what was once expensive has now been made cheap. Exactly the opposite is true.

Before the dam, a float trip down the river through Glen Canyon would cost you a minimum of seven days' time, well within anyone's vacation allotment, and a capital outlay of about forty dollars—the prevailing price of a two-man rubber boat with oars, available at any army-navy surplus store. The rubber boat was the only *special* equipment needed for the voyage. Sleeping bags would be desirable, but then as now almost every vacationer would own one of those beforehand. A life jacket might be useful but not required, for there were no really dangerous rapids in the 150 miles of Glen Canyon. As the name implies, this stretch of the river was in fact so easy and gentle that the trip could be and was made by all sorts of amateurs: by boy scouts, campfire girls, stenographers, little old ladies in inner tubes. Guides, professional boatmen, giant pontoons, outboard motors, radios, rescue equipment—not needed. The Glen Canyon float trip was an adventure anyone could enjoy, on his own, for a cost less than that of spending two days and nights in a Page motel. Even food was there, in the water: the channel catfish were easier to catch and a lot better eating than the striped bass and rainbow trout dumped by the ton into the reservoir these days. And one other thing: at the end of the float trip you still owned your rubber boat, usable for many more such casual and carefree expeditions.

What is the situation now? Float trips are no longer possible. The only way left for the exploration of the reservoir and what remains of Glen Canyon demands the use of a powerboat. Here you have three options: (1) buy your own boat and engine, the necessary auxilary equipment, the fuel to keep it moving, the parts and repairs to keep it running, the permits and licenses required for legal operation, the trailer to transport it; (2) rent a boat at fifteen to seventy dollars a day plus fuel costs; or (3) go on a commercial excursion boat, packed in with other sightseers, following a preplanned itinerary, at a cost of twenty-five dollars per day per person (see Author's Notes).

The inescapable conclusion is that no matter how one attempts to calculate the costs in dollars and cents, a float trip down Glen Canyon was much cheaper than a powerboat tour of the reservoir.

Being less expensive, as well as safer and much easier, the float trip was an adventure open to far more people than will ever be able to afford motorboat excursions in the area now.

What about the "human impact" of motorized use of the Glen Canyon impoundment? We can visualize the floor of the reservoir gradually accumulating not only silt, mud, waterlogged trees and drowned cattle, but also the usual debris that is left behind when the urban, industrial style of recreation is carried into the open country. There is also the problem of human wastes. The waters of the wild river were good to drink but nobody in his senses would drink from Lake Powell. Eventually, as is already the case at Lake Mead, the stagnant waters will become too foul even for swimming. The trouble is that while some boats have what are called "self-contained" heads, the majority do not; most sewage is disposed of by simply pumping it into the water. It will take a while, but long before it becomes a solid mass of mud Lake Powell (Jewel of the Colorado) will enjoy a passing fame as the biggest sewage lagoon in the American Southwest. Most tourists will never be able to afford a boat trip on this reservoir but everybody will be able to smell it.

All of the foregoing would be nothing but a futile exercise in nostalgia (so much water over the dam) if I had nothing constructive and concrete to offer. But I do. As alternate methods of power generation are developed—and Glen Canyon Dam is already plainly obsolete as an efficient power producer—or as the nation establishes a way of life adapted to actual resources and basic needs, so that the demand for electrical power begins to diminish, we can shut down the Glen Canyon power plant, open the diversion tunnels, and drain the reservoir.

This will no doubt expose a drear and hideous scene: immense mud flats and whole plateaus of sodden garbage strewn with dead trees, sunken boats, the skeletons of cattle and long-forgotten, decomposing water-skiers. But to those who find the prospect too appalling I say, give nature a little time. In five years, at most in ten, the sun and wind and storms will cleanse and sterilize the repellent mess. The inevitable floods will soon remove all that does not belong within the canyons. Fresh green willow and tamarisk, box elder and redbud will reappear; and the ancient drowned cottonwoods (noble monuments to themselves) will be replaced by young of their own kind. With the renewal of plant life will come the insects, the birds, the lizards and snakes, the mammals. Within a generation—thirty years—I predict the river and canyons will bear a decent resemblance to their former selves. Within the lifetime of our children Glen Canyon and the living river, heart of the canyonlands, will be restored to us. The wilderness will again belong to God and the people.

# 6. From jeep trails to power plants

When I first saw the land this book is about, I thought like Captain Macomb that a more utterly worthless region could hardly be imagined. In other words, I thought such a beautiful land so full of marvels was fairly safe from the cash-register mentality. Wrong, of course. As the genius of American commerce has discovered, *almost anything can be sold.* Your own mother is merchantable— at least as glue, lampshade covers, a cake of soap. Certainly your home is a saleable item and Utah businessmen, as I have already suggested, have not been laggard in discovering this. But for a while it seemed as if the canyonlands might somehow be overlooked. Wrong.

Even as late as 1948 most of southeast Utah appeared as a blank space on the maps. Terra incognita. All that unbreathed air. A reserve of clean air and unfenced space sufficient for the needs of all 150 million Americans. Wrong. Rough country known only to a few cowboys, a few half-crazy prospectors, a few mostly crazy wanderers like Norman Nevills, Everett Ruess, Frank Wright, Kent Frost, Ken Sleight and about a dozen others I could name. All that untrammeled sunlight through which a human hand might never pass. Wrong. If Paradise is that which existed before men invented evil, then the slickrock and canyon wilderness was indeed a part of earth's original Paradise.

But men did invent evil, especially in the form of ponderous social institutions (such as the nation-state) which in turn invented further evil of so refined a cruelty and gross a magnitude that all connection to the human has been lost. Nazi Germany, for example; and Communist Russia; and then such things as the atomic bomb. Hiroshima, Nagasaki—even Hitler and Stalin were never able to equal what happened there. Governments invented evil on the grand scale and a sweaty scramble began in Utah's canyon country for the then-rare metal uranium. It was found, too, culminating in the big strikes of 1955, and within a few years the canyon country was overlaid with a network of jeep and truck roads and sprinkled here and there with hastily bulldozed landing strips. The assault was under way.

The second phase of the creeping disaster began and ended with the construction of Glen Canyon Dam, perhaps the biggest single act of vandalism ever committed by the government of the United States against the people of the United States. The purpose of the dam was said to be the generation of cash through electricity with which to finance other reclamation projects in the Upper Basin States. The purpose of these other projects, not explained at the time, has turned out to be what might have been expected: subsidizing cheap water supplies for the benefit of a few thousand tax-supported agri-businesses in New Mexico, Colorado, Wyoming and Utah. As an ancillary benefit it was said that the dam would provide a watery horizontal surface on which motorboats could be exercised. A third purpose would be water storage. As a power generator, however, the dam has proved obsolete less than ten years after its completion, unable to compete with the fossil-fuel plants already under construction in the same vicinity. As a recreation utility, the dam merely duplicates what was already available at Lake Mead some 375 miles down-river. As a storage depot the dam costs us more than it saves: it is estimated that the annual loss of water through evaporation alone is enough to supply a city of 25,000 people. The loss through percolation into the highly porous sandstone that surrounds the reservoir is not known. All this for close to a billion dollars, and the lives of nineteen workmen.

These facts were available before the dam was built. Why then *was* it built? We must confess that no one really knows. But like all such shrines to the ineffable and mysterious (the Great Sphinx, for example) it has turned out to be a profitable tourist attraction. People come to Glen Canyon Dam as they do to The House Made Of Beer Bottles in Rhyolite, Nevada—to marvel, to ponder, and to wonder why. But they stay to buy gas, post cards, motel lodging and the marvelously horrible food for which tourist restaurants are famous the world over.

Thus was born motorized mass industrial tourism in southeast Utah. The idea spread like Asiatic flu. If Glen Canyon could be submerged at a profit, why not pave over as much as possible of the surrounding landscape? Asphalt too is known to be a reliable tourist lure.

And so when a portion of the canyon country adjoining the conjunction of the Green and Colorado rivers was set aside by Congress as a national park, presumably to protect it from commercial exploitation, the pressures began at once to open it up to the road builders. For roads, paved roads, are the primary essential in creating an industrial tourist market. Without paved roads you get only the people who want to visit an area for obscure private reasons, probably of an immoral nature. With paved roads, however, you attract as if by magic every American who owns an automobile and knows how to back it out of the garage and how to point it down the highway. The profit lies in heavy volume, as any business-administration major knows.

Efforts were made to save Canyonlands National Park from becoming another "scenic drive" type park, complete with "auto nature trails," but as usual *compromises* had to be made because of political pressure from Utah business interests. A paved road has already been built as far as Squaw Spring; this road will be extended to a point overlooking the confluence of the two rivers, and from there a loop drive will be constructed close to the entrance to Chesler Park, center of The Needles area. This loop drive, in turn, will probably be joined to a road leading south to Natural Bridges National Monument. Another paved road will be built to Grandview Point and Upheaval Dome. Present jeep trails except the one inside Chesler Park will remain in use.

A few words on the word "compromise." Compromise used to mean an agreement whereby two parties to an issue each gave up something in return for something. Nowadays, in conservation controversies, it seems to mean an agreement whereby one party (the conservationists) yield up part of what they wish to save in order not to have to lose it all. That's called a compromise. If the term were strictly applied, it should mean that every time a road is built into a previously unspoiled area, then an equivalent length of road would be closed off, rolled up or otherwise removed from some other area. The same goes for dams, power plants and uranium mills.

After the invasion by the mining industry, the construction of Glen Canyon Dam, and the development of Canyonlands National Park, the rape of the wilderness continued with a comprehensive highway-building program for southeast Utah. Chiefly for the benefit of the trucking companies, a four-lane freeway has been blasted straight through the center of The San Rafael Swell, another totally uninhabited area, at a cost of millions of dollars to us taxpayers. Although no one lives there, this new road will save truckers an hour or so of operating time in hauling bathroom fixtures and power lawn mowers and so forth from Denver to Las Vegas.

Old state road 95, sketched in the first chapter of this book, has now been paved for all but forty of its 130 miles; the remainder is scheduled for completion soon. Two new paved roads have been built from Natural Bridges to Hall's Crossing marina and to Mexican Hat, thus further bisecting and sub-bisecting that once primitive area. The Utah State Highway Department's proposed super-parkway from the Colorado border to the Arizona border has already been mentioned.

All this is merely the beginning. The Four-Corners Regional Commission, representing the Colorado Plateau province of the four adjoining states of Utah, Colorado, New Mexico and Arizona, has presented to Congress a highway-building program calling for 5,699 miles of "high-priority roads" in the Four-Corners region at an estimated cost of $808,924,000. The commission hopes to gain congressional authorization for fifty-one separate highway projects in southern Utah alone, totaling 1,489 miles and costing over $280 million (Highway Department estimates). The purpose of this highway-construction program, as its proponents readily admit, is "economic development"—that is, to bring the benefits of industrialism and urbanism to one of the last-remaining extensive primitive areas

Overhang Arch in Davis Gulch, Escalante area. Monolithic sandstone lacks the intricate and beautiful detail of igneous and metamorphic rock; but in the mass it takes on the most monumental forms.

in the coterminous United States.

You might think this program, if ever put into effect, sufficient in itself to complete the blighting of our Southwest. But even while the road struggle gets under way, with its outcome in doubt, the developers and promoters have begun an assault on still another front: the sky.

The American Southwest was once famous for its pure air, vivid light and dazzling visibility. There was a time, since forgotten, when Eastern physicians would send their serious respiratory cases to Arizona or New Mexico. Thanks to the efforts of regional economic boosters, much of the air of the Southwest has been "brought up to standard"—to use the terminology of the Four-Corners Regional Commission. Our air is just as dirty as your air. Central Arizona, for example, is now a good place to catch a disease known locally as "valley fever," a debilitating lung fungus borne far and wide by the winds in our almost-always dusty air. Visibility also suffers. According to Dr. Otto Franz, an astronomer at the Lowell Observatory near Flagstaff, the air of northern Arizona, once renowned for its clarity, has undergone a light loss of 25 percent since 1962 (see Author's Notes).

The canyon country escaped most of this pollution until recently, except for the haze added to the air by a few potash and uranium mills. Then, with the opening up of a huge coal-burning power plant near Shiprock, New Mexico, the pollution assumed a more serious aspect. Now we are faced with a ring of such power plants, all to burn cheap low-grade coal and all planned to produce power by the megawatts. Along with the new power plants, if permitted, will come the usual strings of transmission lines across the landscape, a railway, a pipeline, and five strip mines. Things are coming up to standard: every hogan owner will soon have a deposit of fly ash on his doorstep—just like Flushing and the Bronx.

Glen Canyon Dam was supposed to produce all the power the Southwest would ever need. Deception: less than a decade later the Bureau of Reclamation, prime instigator of the dam, is now one of the principal agencies involved in promoting the new set of fossil-fuel plants. In fact, two of the proposed plants—Navajo and Kaiparowits— would be built within ten miles of the dam itself; and the filth and acids from their 800-foot stacks—called "beauty tubes" by plant designers—will lay down a film of poisonous soot on the bright and shiny waters of Lake Powell, Jewel of the Colorado.

Construction of these utilities is completed, in progress, or about to begin. The first was the Four Corners Plant near Shiprock, operated by the Arizona Public Service Company. After only six years of operation this plant has become the most notorious single polluter of public air in the entire Southwest. The refuse from its stack can be seen for a radius of 100 miles and has been the object of complaints from citizens as far away as Durango, Colorado, and Albuquerque, New Mexico. American astronauts orbiting the earth were able to identify the Four Corners Plant, thanks to its gaseous garbage, as one of the few man-made objects visible from space.

Soon to begin operation is the Mohave Plant near Davis Dam on the Arizona-California border, owned like the other plants by an interstate consortium of utilities and public agencies (see Author's Notes). Now in the initial phase of construction is the Navajo Plant at Page, Arizona, close to Glen Canyon Dam; this plant, if completed, will have a total installed capacity of 2,310 megawatts, making it one of the biggest in the United States. Others in the project or proposal stage are the San Juan Plant, to be located sixteen miles northwest of Farmington, New Mexico, the Huntington Canyon Plant twenty-nine miles southwest of Price, Utah, and the Kaiparowits Plant twelve miles north of Page. The last-named will be the monster of them all, with a projected ultimate capacity of 5,000 to 6,000 megawatts.

These power plants would half-encircle the canyonlands. The effect of their smokestacks, cooling towers, strip mines, truck roads, waste disposal dumps, pipelines, railways, and transmission lines on the surrounding landscape can easily be imagined. The pollution of the atmosphere would be far worse, an insidious smog of fly ash and sulfur dioxide which will spread as sure as the winds still blow over the entire air space of the canyon country.

The plants are being built around the canyonlands for three reasons: (1) ample source of cheap coal; (2) Utah-Arizona pollution-control standards which are lower than those of California (where most of the power is to go); and (3) a sparse and presumably docile population which, in return for a few hundred jobs, will not object to the pollution of their air and the obscuring of their skies.

As designed, the new plants will have pollution-control standards the same as those of the Four Corners Plant, visible from outer space. This means that the two plants to be built near Page, with a combined generating capacity of at least 7,300 megawatts, will emit on the average some 137 *tons* of fly ash daily. Fly ash, as all urban readers know, is an exceedingly fine grade of dirt—"particulate matter" in the jargon of the engineers. It diffuses through air like a dye through a glass of clear water. The effect on the atmosphere of the canyonlands will be the same. Visibility will be reduced from the customary 60 to 100 miles to more like 15 miles, or little more than the width of the Grand Canyon at Bright Angel Point. Swimmers

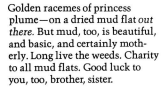

Golden racemes of princess plume—on a dried mud flat *out there.* But mud, too, is beautiful, and basic, and certainly motherly. Long live the weeds. Charity to all mud flats. Good luck to you, too, brother, sister.

and boaters on the Glen Canyon Dam reservoir will plow through a film of soot; hikers will find a coating of the black dust on the ancient monoliths of Rainbow Bridge, the Arches, The Needles—everywhere, *hic et ubique.*

In addition to fly ash the new plants will produce abundant quantities of $SO_2$ (sulfur dioxide) and $NO_2$ (nitrogen dioxide). $SO_2$ when combined with moisture in the air forms sulfuric acid—not healthy for children and other living things. $NO_2$ is the primary ingredient in photochemical smog. The two plants proposed for the Glen Canyon area will produce 735 tons of $SO_2$ and 622 tons of $NO_2$ per day, or about three times the daily production of Los Angeles County. The Navajo Plant alone will also produce 1,840 tons of *recovered* fly ash every day. This waste material, fine as talcum powder, will have to be disposed of somehow, somewhere, in the vicinity of Page and Lake Powell.

Like the network of new highways proposed for the canyon country, these power plants are meant not for current needs but for "anticipated" needs (see Author's Notes). "Planning for growth," it's called. The fact that planning for growth encourages growth, even forces growth, would not be seen as a serious objection by the majority of Utah-Arizona businessmen and government planners. It would merely excite them to greater enthusiasm. They believe in growth. Why? Ask any cancer cell why it believes in growth.

I've been pretty hard on businessmen and bureaucrats in this little essay. At least I've tried to be. My only regret is that I've probably not been hard enough. As a class, as a social power, both in America and in most of the world, they are responsible for the majority (not all) of the difficulties in which we struggle to survive.

Some of these men, when you meet them, turn out to be exactly what you would expect. I know one small businessman in Moab (and sometimes there's nothing smaller than a small businessman) who has lived there for seventeen years and has yet to see Dead Horse Point, Lavender Canyon, Behind-the-Rocks or anything else of interest beyond the town limits. He's been too busy, he says, making money six days a week and going to church twice every Sunday. But he is full of opinions on where new roads should be built (everywhere), who should decide such things (local people only), and who should pay for them (the rest of us).

I wish I could think of this particular fool as typical of them all. It would be convenient and comfortable to have an easily identifiable pack of villains in the picture. But unfortunately I have met others who do not fit the image so obviously. When you sit down and talk with them they tend to resemble real human beings like you and me. They are not stupid or greedy or malevolent; on the contrary most of those I know are decent men, kindly, generous, full of good will. They are also knowledgeable, often sophisticated in such matters as economics, politics, and contemporary events. Whether government official or businessman, they are aware of what is happening in our world. A few of them are friends of mine.

I've accused them as a class of believing in growth for the sake of growth: a blind faith in blind progress. As individuals, however, they know better. They know the meaning of the idea of an optimum point in economic development. Sam Taylor, for example, editor and publisher of the *Moab Times-Independent*, speaks of 10,000 as the "right" population for his community—enough, he thinks, to create a stable economy; not so many as to generate the social evils of a city. (The present population of Moab is 6,000.)

Calvin Black, county commissioner of San Juan County (in which Canyonlands National Park, Natural Bridges National Monument, and the towns of Monticello, Blanding, Bluff and Mexican Hat are situated) is an enthusiastic promoter of any and all kinds of economic development, even the new power plants with their substandard pollution controls—anything for the sake of jobs. Why? "Our biggest export," he likes to say, "is kids and empty pop bottles." He wants jobs so that more of his own six children can find work in San Juan County and live at home.

Nevertheless a man like Calvin Black, too, wants to set limits on quantitative growth. San Juan County now has a population of 10,000; he'll settle for 25,000. The optimum: the children can stay home and bottle their own soda pop. And how many children should his children have? He'll admit that sooner or later the problem will have to be faced but, "it's an individual responsibility." In the meantime he advocates an economy based on mining and manufacturing as well as on the present tourism. Mr. Black is in the mining business.

Across the river in Navajo country the Indians talk the same way. They want jobs, any kind of jobs, where they live now. They do not want to have to move to Los Angeles or Chicago; they want a piece of Los Angeles and Chicago on the reservation. Why? Their reasons are as rationally self-interested as the Calvin Blacks of small-town Utah: so that their children can remain at home. They are not in the least interested in the crazy notion of wilderness preservation.

What shall we say to these people? How can I explain to Indians and county commissioners that it is better to live lean and hungry on the edge of the wilderness than fat in the suburbs of the Gross National Product? How can I help them to understand that bringing that evil plague into the wilderness is the final crime against our American heritage? These are truths they can discover only for themselves.

The tragedy of our national situation lies in the strange historical process by which the majority of us have acquiesced in the creation of a whole that is far worse than the sum of its parts. We have consented to the rule of institutions (presided over by ordinary, well-meaning men whose moral deformity comes chiefly from a total atrophy of the imagination) which serve the worst rather than the best parts of our national character. We have submitted to the domination of an insane, expansionist economy and a brutal technology—*a mad machine*—which will end by destroying not only itself but everything remaining that is clean, whole, beautiful and good in our America.

Unless we find a way to stop it.

Hyde's commentary
on the Escalante wilderness

# 7. On the Escalante wilderness

You come into this land of canyons and mesas from the west, over the Aquarius Plateau from Bryce, past the town of Escalante that has somehow managed to keep a trace of its frontier charm. The Escalante River is so well tucked into the wilderness of slickrock which begins here that it was the last major river in the contiguous United States to be discovered and was named just over a hundred years ago by Powell during his celebrated survey of the Colorado.

The Escalante heads in the snows of the 10,000-foot Aquarius. Flowing to its junction with the Colorado, it cuts through a rock basin thirty miles wide by about sixty long. The basin slopes upward from the remarkable straight cliffs of the Kaiparowits Plateau on the southwest to the Circle Cliffs and the crest of Waterpocket Fold.

This rock basin is formed of sediments in the Glen Canyon series: Navajo, Kayenta, and Wingate sandstones, Chinle shales, and Shinarump conglomerate. The massive Navajo and Wingate sandstones predominating are made up of solidified, cross-bedded sand dunes laid down in the Jurassic time of dinosaurs. Navajo's water-bearing character makes the canyons beautiful as well as livable for many species of creatures, including man. Numerous springs, seeps and side streams provide an unexpected variety of vegetation not found in canyons of the region cut through other formations.

The meanders of the river and its tributaries cut through the soft sandstones to create great bends and overhangs, undercut alcoves, long rock peninsulas.

These are intimate canyons, vast enough to impress yet not so overpowering as to seem unfathomable or out-of-scale with man. They give a sense of joyful anticipation and mystery as you walk the meanders wondering what is around the next bend. The small stream gurgling at your feet, the high, shadowing walls and frequent trees make for pleasant walking and camping. Water and the shelter provided by the great overhangs enticed prehistoric people, too. They lived on the shelves under the overhangs and farmed the sandy bottoms, etched their drawings on selected smooth walls, left a few artifacts, then migrated elsewhere about 1200 A.D.

In the marvelous acoustics of these vaulted chambers the canyon wren's melodious, descending trill flows inward, making the hearer as delighted as a bird to be there listening to the wild music. Walking days are filled with stream crossings. In the warm sun wet feet give pleasant cooling. Walking on sand, your tracks disappear not long after your passing and heighten the impression of wildness the next traveler gets. If you are lucky, you may come upon a recently implanted set of cougar tracks on a wet sandbar, but in a few days or weeks storm or wind will obliterate them, too.

The arches and natural bridges provide a kind of place-name feast for those who need that sort of thing, but walking down a canyon that bends every quarter of a mile or less provides continuous stimulation. After a few such traverses, your eyes begin automatically to scan the scene for its smaller climaxes: an alcove in a blue-black tinted wall; spring-fresh cottonwood foliage against the warm orange-red of a cliff; the patterns of fractures on a sheer wall; brilliant lichen in a sheltered place; a sudden, still pool of water in a dry fork; the intimate, layered waterfalls of Coyote Gulch's side canyons whose high walls end suddenly in an amphitheater gloriously decorated with a plunge pool of crystal water.

This is the Escalante Canyon country which, along with the rest of Glen Canyon, was proposed as Escalante National Park in the middle 1930s and nearly created by a presidential proclamation that just missed getting President Roosevelt's signature. Its wildness has survived three and a half decades of the machine civilization since that fatal omission.

How much longer will it live?

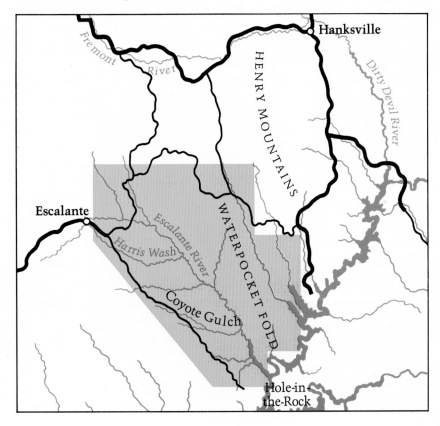

Coyote Gulch, Escalante Wilderness

82

Pool in Scorpion Gulch

Wingate formation, East Moody Canyon

Packer's camp, Escalante River

Junipers, East Moody Canyon

Tributary of Coyote Gulch

Nearing the Narrows of the Escalante

In the Narrows of the Escalante

Concoidal fractures, Coyote Gulch

Slickrock domes near head of Hurricane Wash, Kaiparowits Plateau in distance

Stevens (or Skyline) Arch, Escalante Wilderness

Alcove in wall of the Escalante Canyon

Cottonwoods just leafed-out, Harris Wash

Cliff rose in bloom, Hurricane Gulch

Hyde's commentary
on the Waterpocket Fold country

# 8. On the Waterpocket Fold country

At the northeastern edge of the Escalante basin, the same layers of sediments that provide the material for glorious forms in its canyons suddenly rise in the Circle Cliffs upwarp, then bend and drop steeply. This downward fold extends nearly a hundred miles from the slopes of Thousand Lake Mountain to the Colorado River. The angle of downward fold varies from ten degrees to forty-five. This tilt accounts for many of the features of Waterpocket Fold, including the one that gave rise to its name: the frequent tanks and depressions in the sandstone where runoff collects.

Fine cliffs and beautiful rock forms developed as erosion progressed on the fold. Its best-known feature is Capitol Reef, named for its resemblance to the dome of the nation's capitol. With the reef as its central feature, Capitol Reef National Monument was established in 1937 to include about one-fourth of Waterpocket Fold. In 1968, a presidential proclamation added much of the remaining fold to the monument. An attempt was made in the Ninety-first Congress to nullify this extension, but fortunately that effort failed, and there is now a bill before Congress to confirm the extension and make the monument a national park.

At the northern end of Waterpocket Fold is a wild and beautiful area added to the monument by the 1968 proclamation. This includes the Cathedral Valleys of the South and Middle deserts, several small areas at the head of the fold where fine Indian ruins have been found, and the colorful, rounded badlands of the Bentonite Hills which lie south of the Cathedrals.

South and Middle deserts are dry, nearly parallel valleys just to the east of the fold, with walls and outliers of the pale brown Entrada formation. Where this soft sandstone is capped by the more resistant material of the Curtis formation, it erodes into near-vertical, fluted walls which resemble Gothic forms. Jeep trails made by uranium prospectors and cattlemen wander through the valleys. In many places along the walls are the contrasting black remnants of dikes and sills, now isolated by erosion from the soft materials they long ago intruded. Pinnacles on otherwise flat mesas turn out on closer examination to be volcanic necks, remnants of ancient vent systems that piped the hot lavas to the surface.

The Fremont River, which later becomes Powell's Dirty Devil, is the fold's principal stream. Its beautiful canyon cuts through the fold and is now traversed by Utah Highway 24, enroute to Hanksville. There are a number of small towns along this highway west of Capitol Reef, and except for scattered ranches, these are the only settled places in the general area of the fold.

The southern two-thirds of Waterpocket is wilder still, penetrated only by dirt roads in a few places. One of these crosses the fold near its midpoint on the Burr Trail. Another, coming south from Notom traverses a classic strike valley at the lower edge of the fold in its central section before climbing onto Big Thompson Mesa just to the east. A spur from this road leads to the rim of the mesa for an excellent view of the fold, and a double arch in one of its canyons. At the foot of Big Thompson Mesa, a jeep track leads down the beautiful wash of Hall's Creek.

Because of the steepness of the fold, many of its canyons may never be thoroughly explored. I have started into a number of them only to be stopped by smooth, vertical jumps, or by great chockstones jammed between sheer walls. I remember one in particular whose smooth, sloping slickrock invited walking. It gradually increased in pitch until traction was no longer possible, even with gripping rubber soles. Retreating downward, I was startled at the steepness of the slope I had easily walked up. It was one of those canyons that promised special secrets if you could somehow get past that steep part—and it lures me still to come back and try again.

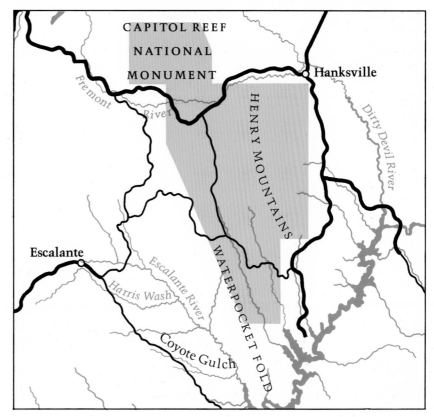

North along Waterpocket Fold, Capitol Reef National Monument

The Red Slide and pedestal rocks, Hall's Creek Wash

Near Cottonwood Creek, Capitol Reef

A "fountain tank" in Waterpocket Fold

Bentonite Hills near Cathedral Valley

Seeps on wall of Hall's Creek Gorge

Yucca and penstemon

Remnants of eroded sandstone, Capitol Reef

110

Mancos shale in strike valley near The Post

Yucca, Capitol Reef

Hall's Creek Gorge

Cathedral Valley, Capitol Reef

# 9. On Canyonlands

Beyond the intimate, watered canyons of the Escalante, beyond the steep sediments of Waterpocket Fold, northeast again just over that great laccolithic swell in Earth's crust that raised the Henry Mountains lie the big, open canyons of an intricately eroded country centered around the coming together of two great rivers. The Green River, flowing southward from its rise in the Wyoming Rockies here meets the Colorado, coming from its sources in the Colorado Rockies.

On the east side of the rivers' confluence, a segment of canyon country has been gathered into Canyonlands National Park. But when the park was established in 1965 certain people thought the original proposal was too big, and some of the wildest and most beautiful parts of the proposal were left out. They should be restored.

As with most of the watershed of the Colorado system, the river is the scenic climax. Cataract Canyon, beginning just below the confluence, is one of the river's wildest sections until it dies in the slack water of "Lake" Powell. Upstream from the confluence, the Green, coming down from the Book Cliffs near Green River, Utah, meanders lazily through magnificently scenic Labyrinth and Stillwater canyons, while the Colorado is flowing down from Moab.

Away from the rivers, this is dry country. Springs and seeps are rare, side streams intermittent and often alkaline. Vegetation is sparse, except on plateau and mountain slopes.

The outside perimeters of the canyon systems are walled by the high, sheer precipices of the Orange Cliffs. The rims of these cliffs provide vistas stretching away for a hundred miles or more on a clear day, as at Dead Horse and Grandview points on the north; at The Needles overlook on the Hatch Point plateau; at North Point, which looks out over The Maze and Standing Rocks.

The uranium craze of the fifties opened this country, lacing it with a network of jeep tracks and seismographic lines. These marks will be on the land for millennia to come. So will some other kinds of marks. A part of the park, curiously, is used as a missile range. The missile shoots began in 1963 and were originally supposed to be over in eighteen months. Yet the program has been going on now

Water Canyon, La Sal Mountains in distance

South Hatch Canyon, Henry Mountains in distance

124

In the La Sal Mountains

Sandbar at Gypsum Canyon

Indian paintbrush, Standing Rocks country

Millard Canyon from rim of Orange Cliffs

Lizard Rock, Standing Rocks country

Pinnacles in the Doll House

Into The Maze

The Fins

Triangle Arch, Lavender Canyon

Early morning in Chesler Park

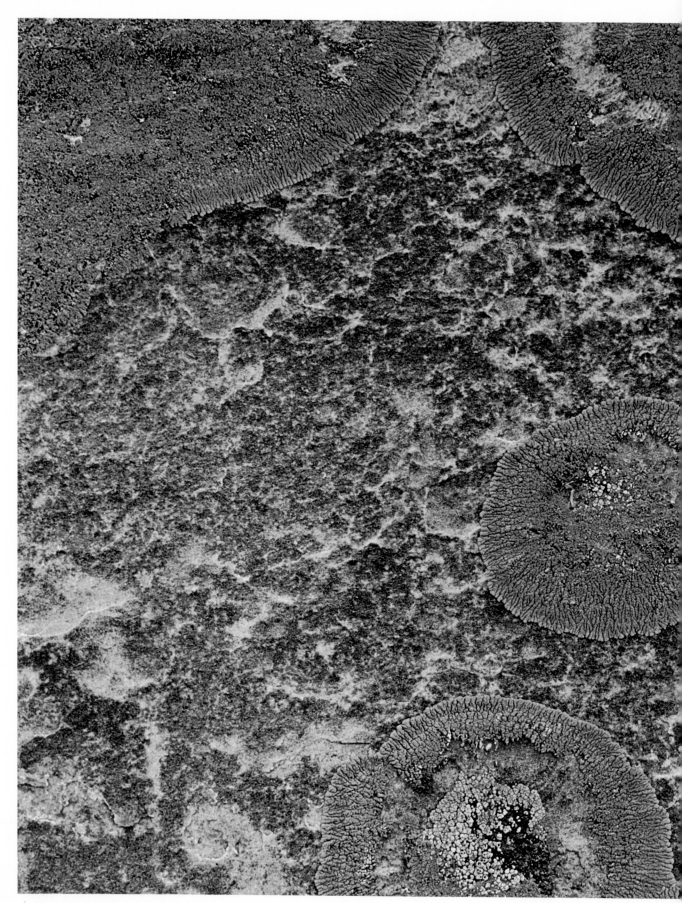

Lichen on sandstone near rim of Water Canyon

# Author's notes

**1. How it was**

Page 19

Everett Ruess, a romantic figure in canyon country history, was a solitary wanderer who disappeared under mysterious circumstances in the Escalante canyon area in the early 1930s. No trace of his body has ever been found. A good picture of his character can be found in his book *On Desert Trails.*

Page 23

As one would expect, any place so pleasant and simple as Natural Bridges National Monument was not allowed to remain that way. It has since been subjected to the industrialized tourism treatment —paved loop roads, paved foot trails, plumbing and electricity, a whole staff of clerks and administrators—and except for the scenery now pushed into the background (suitable for framing) it exactly resembles any other National Park Service "facility."

**2. A history: natural and otherwise**

Page 33

For my version of the geology of the canyon country I have leaned heavily on Dutton, Gilbert and Gregory. (See Bibliography.)

Page 39

The Division of Wildlife Services was formerly called the Predator Control Agency; some government public relations expert thought up the new name. Practices remain the same but the name has been sweetened. Characteristic.

**3. Days and nights in Old Pariah**

Page 46

My information about the history of Paria comes from *Utah, The Incredible Land,* a very handy and nicely illustrated guidebook to the state by Ward J. Roylance.

Page 51

Detailed information about hiking Paria Canyon is available at the Lee's Ferry ranger station and at the BLM office in St. George, Utah.

**4. Fun and games on the Escalante**

Page 53

The cattle industry ranks below tourism, state and federal government work (schools, welfare, Forest Service, BLM, Park Service, highways), mining and general trade as a provider of employment to local citizens. According to an editorial in the *Salt Lake City Tribune*, November 30, 1970, only 22 percent of Utah's cattle industry depends upon the public lands for forage.

Page 54

I recommend a visit to Glen Canyon City. It provides an instructive example of free enterprise at work. I would hesitate to say, however, that it is any uglier than such official government towns as Page. What we have here are two widely different but equally disheartening varieties of civic ugliness. Both are derived, I suppose, from the complete indifference to environmental character and personal human values which is such a fundamental trait of contemporary industrial states.

Pages 55-56

These documents are available on request from the Escalante Chamber of Commerce, Escalante, Utah.

Page 55

"From the residents of eastern Garfield County..." The town of Escalante is in the western half of Garfield County. Nobody lives in eastern Garfield County—nobody at all. As for cattle grazing, it would *not* be affected by wilderness classification under BLM administration, except that motorized operations would be prohibited. Otherwise, cattle grazing would continue.

Pages 57-58

The letter by Nancy E. Williamson appeared in *The New York Times*, Sunday, August 3, 1969.

Page 58

Ken Sleight, Green River, Utah, *Special Report,* privately printed and distributed, May, 1969.

Page 58

For further and current information contact the Escalante Wilderness Committee, P.O. Box 8032, Salt Lake City, Utah 84108.

**5. The damnation of a canyon**

Page 67

Not all of the side canyons have yet been choked by debris. Apparently the wind blows most floating materials into compact masses and tucks them into sheltered corners. Eventually, however, the clogging effect will become general.

Page 68

I have not been able to obtain reliable estimates on the life of Lake Powell and its various marinas. Because of the dams built farther up the river, the siltation process will take longer than expected. A century seems a reasonable guess.

Page 68

Prices quoted here are from 1967; they are no doubt higher now.

Page 68

For example, on powerboat operating costs: the 100-mile trip from Wahweap to Rainbow Bridge and back requires about fifty gallons of fuel for the average powerboat. This means an expenditure, for fuel alone, of at least twenty dollars. The river trip cost nothing, not a cent, and was absolutely headache-free.

**Addendum**

"On November 4, 1970, there was filed a complaint in the District Court at Washington, D.C. by Friends of the Earth, The Wasatch Mountain Club Inc., and Kenneth G. Sleight, Plaintiffs, against Ellis L. Armstrong, Commissioner of Reclamation, and Walter J. Hickel, Secretary of the Interior.

"In the Prayer for Relief the Plaintiffs ask the Court to order the Defendants to perform their statutory duties and take adequate protective measures to preclude impairment of Rainbow Bridge National Monument and to prevent Glen Canyon Reservoir from entering the boundaries of the Monument.

"The Defendants are given 60 days to answer the complaint." (From a report by the Commissioner, Utah Board of Water Resources.)

Sections 1 and 3 of the Colorado River Storage Act (which authorized construction of Glen Canyon Dam) provide as follows:

Section 1—
"..That as part of the Glen Canyon Unit, the Secretary of the Interior shall take adequate protective measures to preclude impairment of Rainbow Bridge National Monument."
Section 3—
"It is the intention of Congress that no dam or reservoir constructed under the authorization of this Act shall be within any national park or monument."

At the time of this writing (12/31/70) the Glen Canyon Reservoir had reached a water-surface elevation of 3,602 feet above sea level. When and if the water reaches the 3,607 point it will begin backing across the boundary of Rainbow Bridge National Monument. When and if the reservoir attains its maximum projected level of 3,700 feet it will have backed water underneath Rainbow Bridge itself to a depth of 46 feet. At that depth motorboats will be able to operate underneath the bridge and even some distance up-canyon from it.

Would that constitute "impairment?" It all depends on one's point of view. Should automobiles be allowed inside the corridors of St. Peter's? Of the Cathedral at Chartres? It's a matter of taste.

In any case, the people who own and operate Utah, Colorado, Wyoming and New Mexico—the Upper Basin states—are alarmed by this lawsuit. The Utah Water Commissioner's report on the matter, partly quoted above, goes on to say, "At a meeting in Denver on December 10, 1970, the combined engineering and legal committees [ of the four states] agreed that the situation could be very serious and demands aggressive action on the part of the states..."

**6. From jeep trails to power plants**
Page 72
Data concerning the Four Corners road program is from the *Congressional Record,* as presented by L. R. Mecham, federal co-

chairman, Four Corners Regional Commission, Department of Commerce, on July 15, 1970.

Page 74

"Valley fever" or coccidioidomycosis, associated with dust and dry air, results from the disturbance of desert soils by agriculture.

Page 74

Statement made by Dr. Franz at an Arizona State Board of Health hearing on May 1, 1970, concerning air pollution problems in northern Arizona.

Page 74

The term "beauty tubes" was coined by Mr. L. M. Alexander, associate general manager, Salt River Project, one of the agencies backing the power-plant projects, in a speech at Page, Arizona, on January 19, 1970.

Page 74

The consortium consists of the Arizona Public Service Company, the Southern California Edison Company, the Public Service Company of New Mexico, The Tucson Gas & Electric Company, The El Paso Electric Company, the Nevada Power Company, the Utah Power & Light Company and the San Diego Gas & Electric Company; also, the Salt River Project (a quasi-public authority, similar to TVA, based in Arizona), the Los Angeles Department of Water & Power, and of course the U.S. Bureau of Reclamation.

Pages 75-76

Statistical and technical data from the following sources:

(a) Letter from Joseph E. Obr, Arizona State Department of Public Health, to L. M. Alexander, Salt River Project, dated March 10, 1970.

(b) Public letter distributed by Citizens for Best Environment, Page, Arizona, Fred W. Binnewies, "spokesman," April 7, 1970.

(c) Letters from John R. Bartlitt, design and chemical engineer, and Mike Williams, PhD., aerospace engineer, co-chairmen of the Los Alamos Chapter of New Mexico Citizens for Clean Air and Water, to the City Council, Page, Arizona, dated January 19, 1970, and April 26, 1970.

Page 76

On "anticipated needs," I quote from the speech by L.M. Alexander cited above:

"...Twenty years from now our country will need four times as much electricity as is needed today. But the situation is even more critical for us in this part of the country...The generating capacity of the West and Southwest will have to increase between five and sixfold during the next twenty years..."

Perhaps. Perhaps not. Who really knows, Mr. Alexander? And precisely what do you mean by "need?" How many neon signs do we "need?" How many electrical toothbrushes, blenders, snowmobile factories, electric blankets, air conditioners, motorized lawn mowers, electrical can openers, illuminated freeways, twenty-four-hour television stations, electric carving knives, Christmas decorations, etc. etc. do we really *need*, Mr. Alexander?

# Acknowledgments

**Hyde's:**

In the course of my seven years of travel and photography in slick-rock country, gathering material for this book, many people provided invaluable assistance. Special acknowledgment is due leaders of the Sierra Club who encouraged the project, gave counsel, or arranged for me to accompany club trips in the area, contributing to my understanding of the country and to the building of a collection of photographs from which the book's illustrations are drawn. Among these are David Brower, Jeffrey Ingram, Paul Brooks, trail companions Howard Mitchell and Tris Coffin.

The National Park Service has also been invaluable in gathering information and helping me get into difficult places. Special thanks go to Franklin Wallace, superintendent of Capitol Reef, and members of his staff, Bert Speed and Virgil Olson, and to Bates Wilson, superintendent of Canyonlands, and his staff.

My family and I are especially appreciative of the friendship, knowledge of the country and introduction to much of it that Hildegard and Parker Hamilton of Flagstaff provided on what started out as a joint "vacation" trip and ended up as a highly productive part of the project. The Hamiltons led us into Robbers' Roost country and introduced us to the Ekker family of Robbers' Roost Ranch and Outlaw Trails. We are grateful to the Ekkers for the wealth of lore and feeling for the country they freely gave. The Hamiltons' and Ekkers' cheerful willingness to welcome our two-year-old son, David, on the trip into The Maze made it an unforgettable experience for the Hyde family.

The unflagging determination of June Viavant of Salt Lake City to save the Escalante wilderness has been an inspiration.

In 1970, Reeves Baker of Escalante, packing our gear into Icicle Springs, made possible a wonderful family trip in Coyote and Escalante canyons.

Finally, Sierra Club Books' Editor-in-Chief John Mitchell's unfailing good humor, quick grasp and firm guidance of the project through the editorial and production stages have been vital.

**Abbey's:**

My thanks to the editors and publishers of *Natural History* and *Sage* magazines, in which portions of this book, in somewhat different form, first appeared.

My thanks to Dr. C. Gregory Crampton for his book *Standing Up Country*, on which I relied for my account of the human history of southeast Utah.

My thanks to June Viavant, Jack McLellan, Ken Sleight, Bates Wilson, Art Ekker, Kent Frost, Ron Smith, John Morehead and the Santa Fe Clearinghouse for much valuable information relating to the various conservation matters mentioned in this book.

My thanks to Ingrid, to Ken, to Debris, to Malcolm, to Harvey, to Tom and to a teen-age horse named Sam for much invaluable companionship during several expeditions into the canyonlands.

142

# Bibliography

Abbey, Edward: *Desert Solitaire*, McGraw-Hill, New York, 1968. Canyon country essays on the theme of wilderness and freedom.

Crampton, C. Gregory: *Standing Up Country*, Knopf, New York, 1964. The best general introduction to the history of the canyonlands of Utah and Arizona. Many excellent photographs.

Cummings, Byron: "The Great Natural Bridges of Utah," *National Geographic* magazine, February, 1910. An interesting early account.

Dellenbaugh, Frederick S.: *The Romance of the Colorado River*, Putnam's, New York, 1902. Another early and authentic account of the region.

Dutton, C.E.: *Report on the Geology of the High Plateaus of Utah*, Government Printing Office, Washington, 1880. A geological classic.

Frost, Kent: *My Canyonlands*, Abelard-Schuman, New York, 1971.

Gilbert, G.K.: *Report on the Geology of the Henry Mountains*, Government Printing Office, Washington, 1877.

Gregory, Herbert E.: *The Kaiparowits Region*, Government Printing Office, Washington, 1931.

Hyde, Philip and Jett, Stephen: *Navajo Wildlands: As Long as the Rivers Shall Run*, Sierra Club, San Francisco, 1967.

Klinck, Richard E.: *Land of Room Enough and Time Enough*, University of New Mexico Press, Albuquerque, 1953. A fine evocation of the desert and canyonlands.

Lavender, David: *One Man's West*, Doubleday, New York, 1956. Very interesting accounts of cattle herding in The Needles, among other things. A good book.

Macomb, J.N.: *Report of the Exploring Expedition from Santa Fe, New Mexico, to the Junction of the Grand and Green Rivers*, Government Printing Office, Washington, 1876. One of the primary sources.

Porter, Eliot: *The Place No One Knew: Glen Canyon on the Colorado*, Sierra Club, San Francisco, 1963. A requiem, in splendid photographs and moving words.

Powell, J.W.: *The Canyons of the Colorado*, Government Printing Office, Washington, 1874, and many later editions by various publishers. The best book ever written about the river and the canyons, and their first systematic exploration.

Ruess, Everett: *On Desert Trails with Everett Ruess*, Desert Magazine Press, Palm Desert, 1950. A collection of letters from the romantic vagabond to friends and relatives.

Stegner, Wallace: *Beyond the Hundredth Meridian*, Houghton Mifflin, Boston, 1954. The best book about Powell and his work.

Waters, Frank: *The Colorado*, Rinehart, New York, 1946. A history, natural and human, of the river and its country.

# About the Sierra Club

The Sierra Club, founded in 1892 by John Muir, has devoted itself to the study and protection of the nation's scenic and ecological resources—mountains, wetlands, woodlands, wild shores and rivers, deserts and plains. All club publications are part of the nonprofit effort the club carries on as a public trust. There are more than 40 chapters coast to coast, in Canada, Hawaii and Alaska. Participation is invited in the club's program to enjoy and preserve wilderness and the quality of life everywhere. For membership information and other data on how you can help, please address inquiries to: Sierra Club, 1050 Mills Tower, San Francisco, California 94104.